Diagnosis of

Plant Viruses & Allied Pathogens

By

Prof. Y.S. Ahlawat

Advanced Centre for Plant Virology
Indian Agricultural Research Institute
New Delhi – 110 012

2010

Studium Press (India) Pvt. Ltd.

Diagnosis of

Plant Viruses & Allied Pathogens

ISBN: 978-93-80012-12-4

Published by:

Studium Press (India) Pvt. Ltd.
4735/22, 2nd Floor, Prakash Deep Building
(Near Delhi Medical Association),
Ansari Road, Darya Ganj, New Delhi-110 002
Tel.: 23240257, 65150447; Fax: 91-11-23240273;
jngovil@gmail.com; jngovil@hotmail.com

Printed and bound at Rajkamal Electric Press, Plot No. 2, Phase-IV, Kundli, Haryana.

PREFACE

For the last 35 years, I have been working as a plant virologist, working with virus and virus-like diseases of horticultural and agricultural crops. During the period between 1996 and 2005, it has been my privilege to conduct group and hands-on training programmes under the project on human resource development in the field of plant virology. The persons trained under these programmes were from University faculty including senior level professors, research organizations and quarantine officers. The practical manual was developed during each group training for the purpose of trainees. However, it has been long felt need by plant virologists in the country for a comprehensive publication of the material provided in training manuals by adding the latest information on the subject. I have also been observing for a long time that there is no single book on practical aspects covering the experimental protocols from field diagnosis, transmission by various methods, field spread, purification of plant viruses, antisera production, serodiagnosis, molecular characterization of plant viruses and molecular diagnosis. Therefore, this book has been written to provide an introduction to the studies relating to virus and virus like pathogens. I gained long exposure on virus and virus like diseases working at the Advanced Centre for Plant Virology at Indian Agricultural Research Institute, New Delhi and its Regional stations at Shimla and Kalimpong.

The first chapter of the book provides brief introduction to virus and virus-like pathogens infecting plant species. In the second chapter, symptoms caused by these pathogens on their host(s) have been covered in detail. In third chapter, protocols for transmission of viruses by various methods including by insect and mite vectors have been described. Chapter four carries information and protocols to study physical properties of the virus in crude extract. Next chapter is on isolation and purification of plant viruses by various methods which is followed by electron microscopy to view the virus particles and phytoplasmas by various techniques. In chapter seven, techniques to produce virus specific antibodies in rabbit (polyclonal antibodies), mice (monoclonal antibodies) have been described. Production of antibodies by phage display technique has also been described in this chapter. After describing the production of antibodies, various serological techniques commonly used for detection of plant viruses and phytoplasmas have been discussed in detail.

In chapter 8, a technique to detect viruses with the help of serology and electron microscopy has been described commonly known as immunosorbent electron microscopy. In chapter 9, techniques have been described for detection of virus and virus-like pathogens using nucleic acid of the pathogen such as DNA-DNA hybridization, DNA-RNA hybridization with radioactive and non-radioactive probes, uniplex and multiplex PCRs, different types of PCR and dsRNA technology. In chapter 10, low molecular weight pathogenic RNAs, the viroids have been described and protocols for viroid detection by R-PAGE, nucleic acid hybridization and PCR have been provided. The next chapter has been provided to detail techniques for cloning and sequencing of plant viruses. In chapter 12, an introduction to microarrays has been given and use of microarrays in virus detection has been mentioned. The chapter 13 provided protocols to develop virus-free plants through tissue culture, heat therapy, shoot tip grafting, ovule culture and by the use of genetic engineering. The last chapter of the publication provides latest information on taxonomy of plant viruses.

At the end of each chapter related photographs have been given. In total 90 photographs have been put in this publication and most of them were collected by me personally. The protocol has been given for each and every experiment and only required theoretical background is given in this publication. The selected readings and selected references have been suggested to the readers. At the end, methods have been given to prepare various buffers and reagents required by various experiments mentioned in the book. The conversion tables are also provided for ready use by the workers.

I am fully conscious that in this publication, there may be shortcomings and defects. I therefore, request my colleagues working in plant virology to communicate/suggest the defects, if any, to me so that I could attempt to rectify them in future editions. However, this publication should be of value to the students and teachers of plant virology, plant pathology, horticulture, microbiology and general botany for undertaking various experiments related to their discipline.

Y.S. Ahlawat

ACKNOWLEDGEMENTS

I am indebted to my colleagues Professor Anupam Varma, Drs. R.K.Jain, Padma Ramachandran, V.G. Malathi, N.K. Chakraborty, F.R. Niazi, K.P. Srivastava, Shelly Parveen and R.P. Pant for developing protocols for various experiments which were compiled in practical manuals of different training programmes conducted by me. These valuable inputs of my colleagues and senior Research Fellows helped me to bring out this practical guide.

At last, but not the least, I wish to thank my students, Dr. R.P. Pant, D.K. Ghosh and Dilip Das who helped me in various ways to bring out this book.

ABOUT THE AUTHOR

Yashvir S. Ahlawat (*b* 1943) obtained his M.Sc. and Ph.D degrees from Agra University. He carried his research assignments at Indian Agricultural Research Institute, (IARI) New Delhi. During his research career he discovered several virus and phytoplasma diseases infecting temperate fruit crops, citrus, maize, wheat, Brassicas, radish, vegetables and flowers. However, the major interest of Prof. Ahlawat had been in virus and virus like diseases of citrus. Prof. Ahlawat served as Incharge of advanced centre for Plant Virology developed with the assistance of UNDP of FAO and also the Virology Unit of IARI. He was also assigned the responsibility of Nodal officer to monitor about 100 World Bank funded NATP projects at IARI. During his 37 years of research career, Dr. Ahlawat guided 9 Ph.D and 4 M.Sc. students. He published over 150 research papers in reputed scientific journals. He handled both National and International projects on virus and virus-like pathogens such as: Indo-Frech project on citrus greening, ICAR-AP-Cess Fund on diagnostics, three National Agriculture Technology Projects one each on Diagnostics, HRD, Priority setting, monitoring and evaluation, DBT project on National Facility for Virus diagnosis and quality control of tissue culture-raised plants. Under HRD project Dr. Ahlawat conducted 4 group trainings of 15-20 Scientists and 18 hands- on trainings for 3-6 months on various aspects of plant virology. One refresher course was organized for senior professors from Indian Universities on molecular diagnosis of plant viruses. Based on the quality of training on detection of plant pathogens, ICAR Appreciation Certificate 1998 was awarded to Dr. Ahlawat. The research contributions made by Prof. Ahlawat have been recognized nationally and internationally. Nationally, he received several honors and awards such as Professor M.S.Pavgi Award 2005 from Indian Phytopathological Society (IPS), Purnima Raychaudhuri memorial Lecture Award 2004, H.D. Chapman Award 1999 for outstanding research contributions in Citriculture from Indian Society of Citriculture (ISC). Internationally, he was awarded first James M. and Adeline Wallace Research award 1998 by IOCV, Universty of California. Dr. Ahlawat is a member and fellow of several national and international professional societies and served them in various capacities such as President IPS and several others. As FAO Fellow, Dr Ahlawat worked at the University of California, Riverside and at Instituto Biologico at Sao Paulo, Brazil. He visited University of California again as IBPGR Expert to develop guidelines for safe movement of citrus germplasm. He participated and chaired scientific sessions in several national and international symposia in India and abroad. Professor Ahlawat visited INRA at Bordeux in France twice as visiting Professor. He was consultant to the Royal Govt. of Bhutan and visited Bhutan in 2001 and 2002. He edited 5 books, contributed several book chapters and popular articles. After his superannuation in 2005, he has been working as Emeritus Scientist.

CONTENTS

CHAPTER 1

INTRODUCTION

The microbes especially fungi and bacteria were established as causal organisms of many diseases of plant species. Since 1880 Koch's postulate was considered necessary to establish the pathogenicity of microorganisms. But there were several contagious diseases where no microorganism could be established. For example, Colour breaking in tulips was known for a long time in the Netherlands, but the cause of the abnormality was not known. Although some nurserymen were aware about the secret to the production of colour breaking in flowers of tulip plants (by way of tuber grafting). Other historical examples are variegation in leaves of *Abutilon striatum* and degeneration of potato which were known in Europe since 18th century. The cause of such maladies could not be established till Dr. Martinus Willem Beijerinck in 1895 established that the cause of tobacco mosaic (already established a transmissible disease by Adolf Edward Mayer, 1882-86) was not a microorganism. He described it as *Contagium Vivum Fluidum*. Henceforth, the term virus was used for any such infectious matter where no microorganism could be found.

The number of virus diseases in plants greatly increased after 1920. The criterian for identification of a viral diseases was:

1. Their infectious nature
2. Sub-microscopic in nature (can not be seen in light microscope)
3. Aging at room temperature
4. Treatment with certain chemicals

Between 1894-95 Hashimoto, a Japanese farmer showed that there was a relationship between dwarf disease of rice and a leafhopper. Fukushi however, experimentally established that the leafhopper, *Nephotettix apicalis* was the vector of rice dwarf disease.

Another important discovery was that the tobacco mosaic virus developed local lesions on inoculated leaves of *Nicotiana glutinosa*. The number of local lesions could be correlated with the concentration of the virus in the inoculum (Holmes, 1929). Later, several viruses were shown to have local lesion hosts. The local lesions on a host are still used as a basis for quantitative assays of viruses.

However, proper identification of viruses remained uncertain till a landmark discovery made by Wendell M. Stanley in 1935 from United States of America. He crystallized a protein from mosaic affected tobacco plants and showed that the virus was proteinaceous in nature. This discovery brought Nobel Prize to him. Later Bawden and Pirie demonstrated that viruses are not only proteins but are nucleoproteins (having nucleic acid and proteins).

The viruses were although crystallized but no body could see them in microscope or otherwise. In 1939, German scientists, Kausche, Pfankuch and Rusca observed rod-shaped particles of tobacco mosaic virus in newly constructed electron microscope. The viruses by this time progressed as:

1. Distinction by symptoms
2. Transmission and host range
3. Natural spread
4. Crystallization of viruses
5. Observation of virus particles

With the development of modern electron microscope and ultracentrifuges since 1945 onward several viruses could be isolated from their host. So far about 700 plant virus diseases have been reported. However, plant virology gradually developed as a new discipline from Plant Pathology.

Many discoveries were made between 1950 to 1975 on virus chemistry and structure. A major technical advance in plant virology was the discovery of density gradient centrifugation which helped the purification of viruses. Using density gradient centrifugation, individual components in the virus preparation are sedimented according to their sedimentation coefficients where virus particles are settled as band in the gradient.

Viruses have several features in common with macromolecule of healthy plant and animal cells. The study on Chemistry and Biophysics of such macromolecules is now termed as *Molecular biology*. The researches in this area have provided us more information and knowledge in the areas of virus structure, function and biological behaviour.

The genome of viruses is either ribonucleic acid (RNA) or deoxyribose nucleic (DNA), either one strand or a few segments encapsidated together or separately.

The causal agents of several plant diseases especially known as *yellows* with symptoms like witch's broom, phyllody and proliferation of flowers could not be established as no pathogen could be isolated from plants showing these symptoms. These diseases were transmissible similar to plant virus diseases. Among such diseases, peach yellows is known since 1888 but no causal organism was established. Graft transmission and leafhopper (*Macrosteles fascifrons*) transmission of peach yellows was established by Kunkel (1924, 1926). A breakthrough relating to such diseases came in 1967 by the reports of Doi *et al.* when they visualized some structures in ultrathin sections of mulberry dwarf which were similar to mycoplasmal infections known to cause diseases in animal and humans. These structures were pleomorphic (no definite shape like viruses) in nature and were tentatively called as *mycoplasma-like organisms* (MLOs) and were put under the class, mollicutes. MLOs reacted to antibiotics and were also transmissible by insect vectors. Koch's postulate was fulfilled with some mollicutes like the one associated with citrus stubborn disease where a spiral form organism was cultured in synthetic medium. Based on its helical morphology, this organism was named as *Spiroplasma citri. S. citri* caused typical symptoms of stubborn disease by feeding the innocuous vector, *Euscelis plebejus* to plants. These insects were made innocuous by injecting the culture of *S. citri* and were then fed to healthy plants.

In spite of this, there were still many diseases where MLOs etiology was established but the associated organisms were not cultured like spiroplasmas. Their properties were distinct from typical mycoplasmas of animals. Therefore, based on their genetical distinction, they were later termed as *Phytoplasmas*. The technology of ultrathin sectioning established the cause of many such diseases and fastidious bacteria (citrus greening bacterium) and the detection of *Rickettsia-like bacteria* infecting plants.

Although viruses and phytoplasmas were well established as causal pathogens of plant diseases but still some virus like diseases such as potato spindle tuber and citrus exocortis etc. remained with uncertain etiology. Diener in 1967 reported for the first time that the causal incitant of potato spindle tuber consisted only pathogenic RNA of low molecular weight (110000-130000) and termed it as *viroid*. These pathogens were much smaller than viruses (250-400 nucleotides). After the discovery of viroids, several diseases like citrus exocortis, hop stunt, chrysanthemum stunt, cucumber pale fruit etc. were found to have viroid etiology. One of the characteristic of viroid is to induce stunting in plants which was exploited to develop dwarfed citrus trees for several years in countries like Australia. However, now dwarfing rootstocks are known hence this practice has been abandoned.

CHAPTER 2

SYMPTOMS CAUSED BY VIRUSES AND VIRUS-LIKE PATHOGENS

Viral infections can affect the growth, development and metabolism of an infected host plant. The symptoms which appear on an infected plant are the first diagnosis of the disease which normally referred as "Field diagnosis". The viruses are also named on the type of symptoms they cause on their original host plants. For example, if a tobacco plant is observed with mosaic symptoms, the disease will be called as tobacco mosaic and the causal virus as tobacco mosaic virus. However, now a days such nomination is confusing as similar symptoms may be caused by other disease agents like phytoplasmas, viroids and rickettsia and bacteria-like organisms.

The symptoms in the field are of considerable diagnostic value especially to the growers to avoid seed or plant multiplication from infected plants. The economic value of the virus disease depends on its severity and incidence. For example, foorkey virus disease of large cardamom in Sikkim and Darjeeling hills is so damaging that the growers carry themselves the eradication programme in large cardamom plantations. The per cent incidence of a disease can be calculated by the following formulae:

$$\text{Percent disease} = \frac{\text{Number of plants infected}}{\text{Total number of plants}} \times 100$$

The symptoms of virus diseases to be described in this chapter are both macroscopic and microscopic or more appropriately external and internal symptoms. Usually infected plant will develop one type of symptom but sometimes more than one type symptoms appear by single infection and in such cases, the symptoms are called syndrome. However, if a variety of symptoms develop in a plant due to infection of two or more viruses then it is referred to as synergism or a synergistic effect. The symptoms due to viral infection can appear at any part of the host plant. The symptoms can be grouped as:

LOCAL SYMPTOMS

The primary symptoms develop at the site of virus entry on the inoculated leaves of a host plant are known as local lesions. The lesions may be necrotic (Fig. 1) or chlorotic (Fig. 2). Such symptoms are developed by mechanically transmissible viruses and are very useful for bioassay of the virus. The local lesions may be of the size ranging from pin point to pin head. The lesions in a host plant are developed by hypersensitive reaction of the host. Sometimes, the lesions are not visible clearly but can be seen by clearing the leaf with ethyl alcohol and then stain with iodine. Occasionally, the virus from local lesions may move to uninoculated leaves of the host plant and can develop similar lesions over there causing systemic infection.

SYSTEMIC SYMPTOMS

Systemic symptoms may be external which are shown by an infected plant or they may be internal in which abnormality develops in cells of affected plant. Internal symptoms are observed in light or electron microscope.

EXTERNAL SYMPTOMS

Reduction in growth

Dwarfing or stunting due to viral infections is a common phenomenon. The stunting may be in whole infected plant or only some parts of the plant. The stunted growth is responsible of reduced yield and quality of product *e.g.* foorkey disease of large cardamom (Fig. 3). In addition of viruses, phytoplasma infections do cause stunting effect.

MOSAIC SYMPTOMS

The symptoms of mosaic normally appear on leaves of infected plants but may also appear on stem, fruits and floral parts. Mosaic symptoms are characterized with various shades of dark green, light green and yellow. Cucumber mosaic (Fig. 4) is a good example of mosaic symptoms. If the leaves of infected plant show dark and light green shades, it is called as *mottling e.g.* Canna mottle and citrus greening disease (Fig. 5) but if green and yellow contrast is there, it is referred as *yellow mosaic e.g.* apple mosaic

(Fig. 6). In case the discoloured parts are sharply bordered but circular, then it is called as *flecking* or *spotting* depending on their size (Fig. 7).

CHLOROSIS SYMPTOMS

Chlorosis is a stage when the whole leaf may become chlorotic and mostly associated with "yellow" such as barley yellow dwarf or mungbean yellow mosaic (Fig. 8). Sometimes the interveinal areas become chlorotic but veins remains green (Fig. 9). In other cases chlorosis may be confined to the leaf veins and such symptoms are called as *veinal chlorosis or vein yellowing* (Fig. 10). Sometimes veins become translucent and these symptoms are referred as *vein clearing* (Fig. 11). Sometimes woody plants show chlorosis of veins in a beautiful pattern and are known as *line pattern* (Fig. 12). Whereas, in some cases the chlorosis developed along the veins and such symptoms are called as *vein banding* (Fig.13).

Colour of the leaves may disappear totally due to viral infection and such symptoms are known as *blanching*. Sometimes leaves of infected plants show reddening or purpling due to pigmentation as has been observed with carrot red leaf disease. The brown colour may also develop in leaves as in case of pea early browning disease. There may be bronzing due to superficial necrosis of epidermis.

RING SPOT SYMPTOMS

These are the common symptoms with some viruses. The rings may develop on leaves or even fruits and the surroundings of the rings may be chlorotic to necrotic (Fig. 14). Indian citrus ring spot (ICRSV) and papaya ring spot diseases are good examples of such symptoms.

NECROSIS SYMPTOMS

When necrosis is developed to larger areas and the invaded tissue are killed then the symptoms are called as *leaf necrosis* or *stem necrosis* depending on the part of the plant got affected such as sunflower necrosis and potato stem necrosis etc. (Fig. 15). Necrosis may also occur on fruit or seed. Necrosis may also develop in veins and then it is called as *veinal necrosis*. There may be necrotic streaks on petiole and stem or may be necrosis at the tip or buds, such symptoms are referred as *tip necrosis or bud necrosis*.

MALFORMATION SYMPTOMS

Irregular development of a plant organ is called as *malformation* (Fig.16). *Leaf rolling* (Fig. 17), *leaf curling* (Fig. 18), *leaf distortion and leaf reduction* (Fig. 19) are common with many viruses, phytoplasmas and viroids infections.

Rugocity (retarted growth of veinal tissue) and rosetting are other forms of malformation (Fig. 20). An increase in the number of cells is called as *hypertrophe* but if the cell number of an organ is reduced, it is known as *hyperplasia e.g.* Stem pitting symptoms in citrus due to citrus tristeza (CTV) infection (Fig. 21). However, a reduction in any organ size of infected plant is known as *atrophe*. Necrosis of leaf tissue is also common with viral infections (Fig. 22).

ENATION AND TUMOR SYMPTOMS

Small outgrowths may appear on leaves especially on veins of infected plant which are known as *enations*. Sometimes swelling or blisters are formed on leaves of infected plant (Fig. 23). Larger swellings on stem and roots are called *tumors* (Fig. 24). Pea enation mosaic and maize vein enation are good examples of enations on leaves. Other examples are Fiji disease of sugarcane. Due to infection with clover wound tumor virus, clover plants develop tumors on stem and roots.

SYMPTOMS OF COLOUR DEVIATION IN FLOWERS

Virus induced colour 'break' symptoms in tulip plants are known for a long time. Such symptoms are also common with wall flower and stock plants with infection of turnip mosaic virus. Break in colour results due to loss of anthocyanins. Flowers turn leaf-like and green in greening or *virescence* in cotton due to phytoplasma infection.

SYMPTOMS OF ABNORMALITIES IN POLLEN, SEED AND FRUIT

Mosaic and ring spot symptoms on fruits have already been described earlier. However, there may be several types of abnormalities in fruits of infected plants such as fewer in number, smaller in size, misshapen and changed texture. Cucumber fruits become misshapen and tasteless due to infection of cucumber mosaic virus. Skin texture is changed in apple fruits due to infection with apple rough skin or star cracking viruses (Fig. 25). Seed quality and quantity is also reduced due to virus infections. The virus may transmit through seed as is known in lettuce and in leguminous plants. Pollens may become sterile of tobacco plants infected with cherry leafroll virus. The flower parts may be reduced and there is no seed setting as in onion seed crop due to infection of onion yellow dwarf virus (Fig. 26). Soybean seed become mottled due to soybean mosaic virus.

WILTING SYMPTOMS

Wilting of infected plants is also common with infection of some viruses *e.g.* tomato spotted wilt virus. Wilting is a result due to excessive transpiration or shortage of water for plants. Advanced stages of stem necrosis of potato and sunflower necrosis may also cause wilting. Wilting is irreversible desiccation of tissue and results in death of plants. Death of epidermal cells is referred as *etching*.

SYMPTOMS CAUSED BY PHYTOPLSMAS

Although many symptoms are common in viral and phytoplasmal infections but the latter shows some characteristic symptoms and properties such as:

1 Phytoplasmal infections cause proliferation of buds resulting in *witch's broom* growth of infected plant (Fig. 27).
2. Excessive branching with negatively geotropic habit of affected plants. Apple and citrus *rubbery wood* diseases are good example of such symptoms (Fig. 28).
3. Suppression of floral development; *virescence or greening or phyllody* of floral parts and proliferation of the inflorescence *e.g.* sesamum phyllody (Fig. 29).
4. All the phytoplasmas are restricted to phloem tissue (sieve tubes) and hence *phloem degeneration* is common with these infections.
5. Most of these pathogens are transmitted by leafhoppers or planthoppers but some are transmitted by psyllids *e.g.* Pear decline.
6. Phytoplasmas are sensitive to tetracyclines but not to penicillin.
7. Some of the mollicutes have been cultured in synthetic media and characterized. These cultured mollicutes are known as *spiroplasmas* and were found to be associated with several plant diseases such as citrus stubborn and corn stunt. These organisms have been placed in family Spiroplasmataceae of class Mollicute and have spiral morphology (Fig. 30).

Other disease inciting agents like citrus greening bacterium and Rickettsia-like organisms also produce symptoms similar to viruses and phytoplasmas.

SYMPTOMS CAUSED BY VIROIDS

Viroids are non-encapsidated RNA of low molecular weight and are agents of infectious diseases in many crops like citrus, chrysanthemum, cucumber, coconut, hop, potato, tomato etc. The examples of viroid diseases are; citrus exocortis, hop stunt, chrysanthemum stunt, potato spindle tuber (Fig. 31), coconut cadang-cadang. In citrus, a yellow corky vein disease has been reported from India which is caused by the synergistic effect of citrus exocortis and hop stunt viroids (Fig. 32).

INTERNAL SYMPTOMS

Some of the changes in cells of virus infected plants are common such as break down of chloroplast in mosaic related symptoms. Abnormal division of cells causes enations or tumors. In addition to these, there are some cytological and histological changes that occur in infected cells are visible in light or electron microscope. Several viruses infect cell nuclei and can break down nucleoli or the nucleolus may become of abnormal size. Some of the examples of such infections are southern bean mosaic, tomato bushy stunt, pea enation mosaic and bean golden mosaic. The chloroplsts become colourless, misshapen and clustered with some viral infections. The mitochondria of virus invaded cells become aggregated. The infections commonly cause either reduction or increase in cell numbers and internal cell necrosis. Lignification of xylem, callosing of sieve plates and degeneration or death of phloem cells also occur in infections like grapevine fan leaf and barley yellow dwarf viruses.

PLANT VIRUS INCLUSION BODIES

Inclusion bodies are found in cytoplasm or in nucleus in virus infected plant. These bodies are virus group specific and are useful for virus identification. Readers are referred to the publication of Edwardson and Christie, 1978 for detailed information on this subject.

CYTOPLASMIC INCLUSION BODIES

These inclusion bodies are the aggregate of virus particles or their proteins and are arranged in a definite manner in cytoplasm. They may be fiberous, and paracrystalline inclusions characteristic of rod-shaped viruses. Crystalline inclusions are formed with both isometric (cucumo and nepo viruses) and rod shaped (tobacco mosaic) viruses. The crystals may be smaller and visible only in electron microscope but inclusions with a size of 15-20 µm can be seen in light microscope.

Some inclusion bodies are only proteinaceous in nature and may be amorphous (watermelon mosaic virus) or cylindrical. In cross sections, the cylindrical bodies look like pinwheels in electron microscope and are characteristic of potyviruses (Fig. 33). The proteinaceous inclusions are also found in plants infected with nepo and como viruses.

NUCLEAR INCLUSION BODIES

The inclusion bodies are also found in the nuclei of cells infected with plant viruses. These may be amorphous (bean yellow mosaic virus) or crystalline (alfalfa mosaic virus).

Inclusion bodies in nucleus or cytoplasm are found in crystalline arrays of either virus particles or virus induced proteins. The virus particles found in these arrays are in larger quantities and are known as *viroplasm*.

The inclusion bodies can be seen by viewing free hand sections or epidermal strips in light microscope (with oil immersion). The sections may be stained with a solution of phloxine or methylene blue or a mixture of both at a concentration of 10 g/L in distilled water. Azure A in 2-methaxiethanol mixed with staining in 9:2 ratio with disodium phosphate in distilled water (0.2 mol/L) is better as it can stain nucleic acid and virus crystal can easily be located. For proteinaceous inclusions, calomine orange 2RS and 'Luxol' brilliant green both in 2-methoxyethanol (10 g/L), mixed just before staining in 1:1:8 ratio with water, orange and green. Plastids that obscure small inclusions can be removed by treatment with triton X - 100 for five minutes before staining.

PROTOCOL

1. Take out epidermal peels from ventral surface of leaves.
2. Keep peels in 1:1 solution of phloxine and tripane blue for five minutes.
3. Wash with water to remove excess stain.
4. Mount in glycerine
5. Observe in light microscope.

Nucleus will take purple stain and inclusion bodies appear pink in colour. TMV produces needle shaped and hexagonal crystalline inclusions in epidermal cells, trichome and subsidiary cells of stoma in profuse number.

Cowpea mosaic virus showed amorphous mass in the cytoplasm of epidermal cells.

Dimorphotheca mosaic virus showed bipolar round inclusions attached on either ends of the nuclei in all epidermal cells. Healthy peels showed clear cytoplasm and distinct nucleus.

LATENT INFECTIONS

There are many plant species especially weeds which can harbor viruses without showing any visible symptom. Such infections are known as latent infections. For example, Cucumber mosaic virus remains latent in weed, *Stellaria media*. Sometimes, a susceptible plant to a virus stops showing symptoms after showing symptoms for some time although the virus remains there. This phenomenon is known as "masking" of symptoms. Latent and masking infections can be detected by back inoculation to a susceptible but not sensitive host.

VARIATION IN SYMPTOMS

Disease is a result of interaction between pathogen, host, and environment. The environmental conditions such as temperature, light and nutrition affect mainly the host but not to virus as viruses do not have metabolism of their own and so are less variable. The genetical composition of either the virus or the host may control the susceptibility of symptom production. The terms normally used in relation to infectivity are:

Pathogenicity : It depends on infectivity of the virus (readiness to infect and invade the host or to multiply within it).

Virus Virulence: Vulnerability to disease

Susceptibility: Readyness to accept the presence of the virus

Degree of infection: it depends on virus aggressiveness and severity of symptoms, virus virulence and host sensitivity.

Severity of disease: depend on virus pathogenicity and host predisposition.

The type of symptoms caused by virus and virus-like pathogens with common examples are given below:

Type of symptoms	Example
1. Mosaic	Tobacco mosaic/cucumber mosaic
2. Mottle	Mottling of French bean
3. Variegation	Abutilon mosaic
4. Vein flecking	Citrus tristeza
5. Vein banding	Apple mosic (intermediate strain)
6. Vein yellowing	Bhendi (okra) yellow net vein
7. Mosaic on stem	Papaya ringspot
8. Colour breaking	Bean yellow mosaic virus in gladiolus
9. Virescence	Cotton virescence
10. Green patches on cucumber fruits	On cucumber fruits due to CMV infection
11. Water soaked rings	On papaya fruits due to papaya ring spot
12. Specks and mottle in seed	Soybean mosaic
13. Necrotic spots on leaves	TMV on *Nicotiana glutinosa*
14. Yellow rings on leaves	Indian citrus ringspot
15. Necrotic ring spot on fruit	Plum pox
16. Fern leaf or shoe string	Papaya ringspot
17. Enation on leaves	Cotton leaf curl, Maize vein enation

18. Blisters on leaves	Turnip mosaic on *Brassica juncea*
19. Cupping of leaves	CMV on sponge gourd
20. Curling of leaves	Tomato leaf curl
21. Twisting leaf and petiole	Papaya leaf curl
22. Leaf roll	Potato leaf roll
23. Reduction in leaf size & shape	Papaya ringspot/ Little leaf of brinjal
24. Proliferation of lateral buds	Witches broom of lime
25. Bushy appearance of plants	Banana bunchy top, Foorkey disease of large cardamom, sugarcane grassy shoot
26. Malformation of leaves	Citrus exocortis viroid on Etrog citron
27. Stem pitting	Citrus tristeza
28. Bark scaling	Citrus exocortis
29. Phyllody of flowers	Sesamum phyllody
30. Elasticity in branches	Apple rubbery wood

CHAPTER 3

TRANSMISSION

Many plants carry viruses but did not show obvious or visible symptoms. In such cases the presence or absence of the virus can be confirmed by grafting scion from suspected plant to a susceptible plant which will display prominent symptoms. Similarly, some plant diseases may be confused with genetic variabilities, physiological disorders such as nutritional deficiency, these maladies can also be differentiated from virus diseases by way of grafting. If the malady is the effect of viral infection, it must be transmissible otherwise not. This method of transmission is used only with dicot plants.

GRAFTING

Various types of grafting are known to propagate horticultural plants such as wedge, cleft, bark, tongue, bridge, inarch grafting etc. but wedge grafting is normally used for virus transmission. Chip grafting is normally used to transmit phloem limited viruses and phytoplasmas in fruit trees. Core grafting is used to transmit a virus from tuber to tuber. Natural root grafting is also known for natural spread of apple mosaic virus. The grafting usually used for virus transmission are detailed below:

Grafting is normally done during winter and early spring when both scion and rootstock are dormant. Grafted plants may be kept in glasshouse to avoid transmission of unknown viruses by insect vectors.

COLLECTION OF SCION FROM FIELD AFFECTED PLANTS/TREES

Best quality scion is obtained from systemically infected shoots of previous year growth in perennials but a symptomatic scion from annuals can be taken during their growth season. Scion should be severed with sharp, clean shears or knives and placed immediately in moistened plastic bags. The operational cutting tools should be cleaned after cutting each scion. This can be done by immersing them in a sterilizing solution such as isopropyl alcohol or with a mixture of household bleach and water (1:9).

For best results, harvested scion should be grafted the same day on good quality and healthy stock plants otherwise store them in a refrigerator. If large quantities of scions are to be harvested during a disease survey then follow these steps:

- Cut all scions to a uniform length, keep their basal ends together and tie them in bundles of known quantity.
- Label them, recording the cultivar, date of harvest, location and symptoms on plant.
- Wrap the base of bundles in moist moss, if available, and put them in polythene bags and seal them to avoid evaporation.
- Store the bundles for short periods, if necessary, in a refrigerator. Avoid freezing.

In grafting as well as budding, the vascular cambium of the scion or bud must be aligned with vascular cambium of rootstock. This vascular cambium initiates callus tissue at the graft and bud unions in addition to stimulate tissue growth on the basal ends of many vegetative cuttings before they have rooted. A cross section of woody plant stem is given in Fig. 34.

Several types of grafts are used for virus transmission but the commonly used are:

CLEFT GRAFT

One of the simplest and popular forms of grafting, cleft grafting (Fig. 35) is a method normally used for transmission of viruses of fruit trees (apple, cherry, pear and peach). The rootstock used for cleft grafting should be at least one inch in diameter and should be straight. The scion should be about ¼ inch in diameter and should have at least three buds. Two scion may be used if the rootstock is thick enough (Fig. 35). The rootstock should be given a smooth cut perpendicular to the main axis of the stem to be grafted. Make a split or 'cleft' through the centre of the stock and down 2-3 inches. The basal end of scions is tapered on both sides. In cleft grafting one scion is usually inserted at each end of the cleft so prepare two scions for each graft which will help to transmit the virus even if it is unevenly distributed in original host. Insert scion on each end of cleft with the wider side of the wedge facing outwords. The cambium of scion should contact the cambium of the rootstock so that virus from scion moves to rootstock. Thoroughly seal cut surfaces with grafting wax or grafting paint.

WEDGE GRAFTING

The scions are collected from infected plants. The cut end of each scion is gently made into a wedge. The stem of healthy rootstock is cut with the help of sharp razor and a vertical incision is made. The wedge shaped scion is gently inserted in the incision of stock and tied with grafting tape or polythene strips. The grafting plants are kept covered for 3-4 days with a polythene bag or may be kept in a moist chamber. The grafted pants should be maintained in glasshouse. The symptoms of the disease will appear on the stock when new leaves develop. The method is used for both annual and perennial plants.

CHIP GRAFTING

A chip of bark is cut from a symptomatic twig. The wood from chip is removed. A similar size chip is cut from the stem of a healthy plant of the same plant species. The chip from infected plant is put in to healthy plant and tied with polythene strips. This method is sometimes used for viruses which are restricted to phloem tissues.

SPLICE GRAFTING

It is normally used to transmit viruses from herbaceous plants. In splice graft, both stock and scion must be of the same diameter. Cut off the stock using a diagonal cut ¾ to 1 inch long. Make the similar cut at the base of scion. Fit the scion to the stock (Fig. 36). Wrap the joint with a rubber grafting strip. Maintain the grafted plants in the glasshouse and record for the development of symptoms on the leaves of stock.

CORE OR TUBER GRAFTING

This type of grafting is used for transmission of viruses in tuber crops. With the help of a cork borer a piece is removed from an infected tuber. The piece should have one or two eyes. A similar piece is also removed from the healthy tuber. The piece of infected tuber is inserted in the cut of healthy tuber. The cuts are sealed with wax. The plants develop from healthy eyes are monitored for symptom development. This method is used to transmit potato viruses.

LEAF PATCH GRAFTING

It is like bark patch grafting. Leaf patches from symptomatic leaf are inserted in to the stem (below the bark) of healthy plants at 3-4 places and tied with polythene strips. The grafted plans are maintained in glasshouse. The virus from patch of infected leaves will travel to healthy plant and the plant may show the typical symptom of the disease. The method is sometimes used to transmit fruit tree viruses.

TRANSMISSION BY BUDDING

Budding is also a grafting technique in which a single bud from the desired scion (virus-infected) is used rather than an entire scion containing many buds. Budding requires the same precautions as grafting. Be sure that the bud and rootstock are compatible. Be careful to prevent drying of buds. Generally viruses from fruit and shaded trees are transmitted by bud grafting. Rootstocks can be grown in the glasshouse. Leaves and side branches of the rootstock should be removed. The stem is cleaned by rubbing it gently with a piece of soft cloth. Mature buds should be used for grafting. Normally T-budding is used for transmitting plant viruses. The method is described below.

In T-budding, budding knives usually with a curved tip are used. First insert the point of the knife and use a single motion to cut the top of T on stem of rootstock then without removing the knife, twist it perpendicularly to the original cut and rock the blade horizontally down the stem to make the vertical slit of the T. Pop open the flaps of the cut and make it easier to insert the bud. The bud to be inserted should have a very thin layer of wood. Insert the bud into the T flaps of the rootstock and slide it down to ensure that it makes intimate contact with the rootstock. Wrap the portion of the graft tightly but do not cover the bud (Fig. 37).

MECHANICAL OR SAP TRANSMISSION

For successful mechanical transmission, minor injuries on leaf surface is necessary which may be done by dusting an abrasive like carborundum on leaves before inoculation. Carborundum powder is a very fine particles of silicon carbide and gives required injury to leaf if rubbed lightly. The abrasive may also be mixed with the inoculum. The following protocol should be used for sap inoculation.

1. The symptomatic young leaves should be collected from a plant infected with a virus.
2. Wash the leaves and gently dry with tissue/blotting paper.
3. Grind the leaves in chilled mortar and pestle or in liquid nitrogen. Use a suitable buffer as grinding medium to avoid loss of virus infectivity.

Suitable buffer must be selected by assaying various buffers on a local lesion host. Normally phosphate buffer of 0.1 molarity with pH between 7.0-7.4 is used but if the sap is acidic pH at 8.0-8.5 must be used. 1% sodium sulphate (reducing agent) may also be added to the buffer to prevent oxidation which loss the infectivity of the virus (for preparation of buffers see Appendix IV).

4. After grinding, pass the leaf pulp through double layer of muslin cloth.
5. For standard extract, grind equal weight of leaves in equal quantity of buffer. It is advisable to use standard extract for sap transmission. The standard extract can be prepared by adding 1 ml of distilled water for each gram of wet weight of diseased tissue used for extraction of sap.
6. Dust 300-600 mesh carborundum or celite on leaves of test plants.
7. Gently rub the cotton pad soaked in standard extract on to carborundum dusted leaves supporting them from lower surface with forefinger or a piece of cardboard. The inoculation should be made only on upper surface of leaves starting from petiole side to the apex of leaf.
8. Wash the inoculated plants immediately after inoculation with the help of a wash bottle to remove excessive inocula and extraneous particles.
9. Maintain inoculated plants in glasshouse or insect-free greenhouse and observe for symptoms development daily.

TRANSMISSION BY DODDER

It is difficult for some plant species to make a successful graft or graft can not be made between two unrelated species which is sometime required for experimentation. Dodder (*Cuscuta* species), a parasitic plant is helpful to establish vascular contacts between two unrelated plant species and hence found to be a good transmitter of some viruses. Almost all the phytoplasmas and fastidious bacteria are transmitted by dodder from their natural host to periwinkle (*Catharanthus roseus*). *Cuscuta* species *viz; C. campestris, C. subinclusa, C. chinensis* and *C. hyalina* have been reported the transmitter of virus and virus-like pathogens.

To conduct dodder transmission, dodder is allowed to parasitize the infected host plant and left there for 2-4 weeks to make enough haustorial connections. The dodder is always twined anticlock-wise for its quick establishment. After the dodder is well established, the transmission experiment can be conducted by two ways:

1. Remove the dodder threads from infected host plant, twin them on healthy test plants anticlock-wise, leave the dodder on test plants for 2-3 weeks and observe symptoms. The experiment should be conducted under insect-free conditions.

2. Healthy test plants are kept around the infected plant carrying dodder. The dodder threads are twined to test plants nearby thus forming a bridge between diseased and healthy plants (Fig. 38). The virus from infected plant will move to healthy plants through the bridge of dodder.

The transmission is more efficient if the pathogen is multiplied within dodder. The dodder should be destroyed after the experiments to avoid parasitism to other plants in the glasshouse.

TRANSMISSION THROUGH VEGETATIVE PROPAGULES

Viruses are systemic in their host plants and hence the organs such as bulbs, corms, roots, runners, suckers and budwood are also infected and give rise to infected plants when used for propagation. These infected vegetative propagules also help in perpetuation and wider dissemination of viruses. These organs are sometimes used to maintain the virus culture for longer periods.

TRANSMISSION THROUGH SEED AND POLLEN

There are over 100 viruses which are known to be transmitted through true seed such as tomato mosaic virus in tomato seed, cucumber mosaic virus in wild cucumber seed, lettuce mosaic virus in lettuce seed and many more. Transmission through seed is mainly through embryo of contaminated seed.

To conduct an experiment on seed transmission of a virus, large quantity (1000-2000) seeds are collected from naturally infected plants. These seeds are sown in seedling trays in the glasshouse. The seedlings are observed for disease symptoms for a period of about one month for herbaceous plant species and for about 6 months for perennial plants.

For pollen transmission, pollens are collected from infected plants and used for pollination to healthy plants. After pollination, the plants are observed for symptoms or the seeds are collected from pollinated plants and used for seed transmission as described above. Viruses infecting *Prunus* species are normally pollen transmitted.

INSECT AND MITE TRANSMISSION

Transmission through mite and insects is a natural and main method of virus spread. About 400 species of animal kingdom are reported to transmit plant viruses. The major vectors of viruses are from the Phylum Arthropoda (94%). Other important vectors (about 6%) are nematodes (Phylum Nematoda). The most important vectors of plant viruses are aphids, leafhoppers, planthoppers, treehoppers, whiteflies, mealybugs, beetles, and thrips. Psyllids are known to be the vector of some phytoplasmas and fastidious bacteria. Some of the insect vectors are shown in Fig. 40.

Mites especially eriophyid mites are known the vector of some plant viruses. The transmission of viruses through mite is circular (virus circulates within the body of mite vector). Pigeonpea sterility mosaic virus, citrus leprosies, peach mosaic and coffee ring spot virus are some of the examples of mite transmitted viruses.

For insect transmission, healthy colonies of insects are raised and maintained on their favorable host plants in the insectory.

APHID TRANSMISSION

The following terms are frequently used during transmission by insects;

1. **Pre-acquisition fasting:** The fasting given to insects before feeding them on infected plant during transmission process. Fasting helps the insects to feed quickly on infected source and acquire the virus.
2. **Acquisition feeding time or acquisition access time:** The time which is allowed to the insects to feed on diseased source.
3. **Incubation period in vector:** The time by which the insect becomes viruliferous or capable to transmit the virus.
4. **Inoculation or transmission feeding period:** The period by which the insects are allowed to feed on test plants after acquisition feeding.
5. **Post-acquision fasting:** The time allowed to insects between acquisition and transmission feeding periods.

The following methods have been used for virus transmission by aphids:

A. NON-PERSISTENT TRANSMISSION

For most of the viruses transmitted by aphids, following protocol is used:

1. Collect aphids from healthy colonies maintained in insectory in a Petri dish by camel hair brush, slightly disturbing them before picking.
2. Give pre-acquisition fasting for 60-90 minutes in Petri dish itself by just covering the dish with muslin cloth.
3. With the help of a camel hair brush, transfer the fasting aphids to virus infected plant or put a symptomatic young leaf in the Petri dish itself for acquisition feeding (Fig. 39). Give feeding only for a few minutes or seconds.
4. Release the aphids on test plants in a group of 5-10 aphids per plant for transmission feeding.
5. Maintain the aphids on test plants for 24 h under chimney tied with muslin cloth at the top or in transmission cages (Fig. 42).

6. Spray insecticide on test plants after 24 h transmission feeding to kill the aphids.
7. Maintain the test plants in glasshouse and observe for development of symptoms.

B. PERSISTENT OR CIRCULATIVE TRANSMISSION

1. Pre-acquisition fasting not required.
2. Give long acquisition feeding time (6-24 h).
3. Give incubation period to aphids after acquisition access of 12 h or more to make the aphids viruliferous(ready for transmission).
4. Transfer viruliferous aphids to test plants and shift them to new test plants after every 24 h till they remain alive.

The virus is retained by aphids even after moulting in case of circulative viruses. This type of transmission is called as *transstadial transmission.* Persistent viruses either do not multiply in vector *e.g.* Barley yellow dwarf virus or the virus multiplies in the vector such as lettuce necrotic yellows. The latter types are known as *Propagative viruses*. Some of the viruses even transmitted through eggs of the insect vector means the next generation of such insects will be viruliferous without acquiring the virus again. Such transmission is referred as transovarial transmission *e.g.* Sowthistle yellow vein virus in its vector, *Hyperomyzus lectucae*.

C. SEMI-PERSISTENT TRANSMISSION

This type of transmission is between non-persistent and persistent transmission.

1. There is no effect of pre-acquisition fasting.
2. Give acquisition access on infected source for 30 min to several hours.
3. After 3-4 days incubation, transfer the insect serially on test plants or even after the acquisition access transfer serially to test plants by transferring after every 24 h. for a week and then destroy the insects by spraying insecticide.

The three types of transmission of the viruses falling in a particular category can be determined while transferring the insects on serial plants after every 24 h as shown in the Table 1.

It would be seen from table- 1 that non-persistent viruses infect only first plant in serial transfer (sometimes two may be infected) and after that no infection was observed. This indicates that the virus was completely lost after infection to one or two plants in serial transfers. Like non-persistent viruses semi-persistent viruses do not infect a few plants in the

Table 1. Determination of different types of transmission by insects

Type of transmission	Plant Number											
	1	2	3	4	5	6	7	8	9	10	11	12
Non- persistent	+											
Semi-persistent					+	+	+	+				
Persistent							+	+	+	+	+	+

+ = plants infected in serial transfer

beginning but infected plants starting from 5th day to 8th day (or plant numbers 5,6,7,8) and after that no transmission was observed. The results indicated that in semi-persistent viruses, an incubation of 3-4 days is required in the vector and then it will transmit for a few days and after that the virus is lost from the insect resulting no transmission. Persistent transmission starts after 6 days or more after acquisition access and once the vector starts transmitting the virus it will go on transmitting through out its life span as shown in the above table that the transmission was obtained up to the last plant in the series.

Only apterous forms of aphid must be used in experimental transmission studies as these forms are convenient to use (alate or winged forms are responsible for natural spread of viruses and difficult to handle for experiments).

TRANSMISSION BY LEAFHOPPERS, PLANTHOPPERS AND TREEHOPPERS

More than 40 species of leafhoppers are known to transmit plant viruses, phytoplasmas and bacteria-like organisms. Leafhopper transmitted viruses are persistent or circulative in nature. The virus undergoes an incubation period in the leafhopper vector for a week or so after acquisition access. After the incubation period is over, the leafhoppers become viruliferous. Some leafhopper transmitted viruses are transmitted in a semi-persistent way such as rice tungro virus transmitted by leafhopper vector, *Nephotettix virescens*. 22 species of plant hoppers and one species of treehopper have been reported as vectors of plant viruses. The only example of treehopper transmission is pseudo-curly top disease of tomato by its vector, *Micrutalis malleifera*.

For experimental purposes, colonies of leafhoppers are raised on suitable hosts in wooden cages in insectory as detailed below:

1. Remove non-viruliferous leafhoppers by a catcher (Fig. 42). One end of this tube is covered by muslin cloth and inserted into a plastic tube with slightly bigger in diameter and of about 75 cm long. The insects are sucked from the open end of the plastic tube.

2. The insects thus collected are released on infected plant in a cage for acquisition access feeding. The cage is (made up of polymethyl methacrylate) covered at one end by nylon mesh or muslin cloth. These cages are of two sizes; having diameter of 2.54 cm and 7.62. The body of the cages is provided with air vents on opposite sides being covered with 60 mesh nylon net and on the top too for ventilation and prevent accumulation of moisture (Fig. 40). A infected plant in a pot is covered by the cage before releasing the leafhoppers. The insects are kept on infected plant for desired acquisition access period (6 h to 4 days depending on the virus).
3. After the acquisition access period, the leafhoppers are collected from infected plant and released on individual test plants which are also covered by cages. Although a single viruliferous leafhopper can transmit the virus but to get good percentage of transmission 3-5 leafhoppers per plant should be released and kept there for 4-6 days and then kill them by spraying insecticide.
4. To determine the survival of the virus in the vector, the leafhoppers should be transferred after every 24 h to new test plant till the insects are alive (serial transfer).

All operations of leafhopper handling are done in a closed chamber called transfer chamber. It is a rectangular wooden chamber measuring 2' X 4' X 1.4' in size having a sliding door of two panels on the front (Fig. 41). By careful manipulation of the sliding glass panels, enough room is made for inserting of hand for removal of cages and then to suck the hoppers by catcher.

5. Keep all the test plants in the glasshouse and record the symptoms appearing on leafhopper inoculated test plants.

WHITEFLY TRANSMISSION

In tropics and subtropics, whitefly transmitted viruses are of high economic importance. In India, cotton leaf curl, papaya leaf curl, cassava mosaic, moongbean yellow mosaic, tomato leaf curl, bhendi (okra) yellow net vein etc. are important virus diseases transmitted by whitefly, *Bemisia tabaci*.

The transmission test by whiteflies is carried out in special type of cages (Fig. 40). These cages are made up by polymethyle methacrylate tube of 1-2" diameter and about 6-8" in length with a hole at the side for releasing whiteflies. The tube is covered at the lower side with the help of muslin cloth. The upper side opening is closed with a light piece of cardboard fitted with a spring. For undertaking the transmission test, the cage may be tied with a bamboo stick close to the plant.

The symptomatic virus infected leaf is inserted between the mouth of the tube and the cardboard by loosing the spring slightly by hand. The whiteflies are reared through the side opening and after the release the opening is closed with a small piece of cotton. The whiteflies thus released will feed on the leaf of desired plant.

The acquisition access period may be few minutes to few hours depending on the virus. The whitefly transmitted viruses are semi-persistent type. Hence the vector may require few hours (4-8 h) incubation before becoming viruliferous. The following protocol may be adopted.

1. The whiteflies are raised on tobacco or any other suitable host in cages similar to the ones used for leafhoppers.
2. Collect healthy whiteflies by an insect collecter in a test tube. Cover the test tube with muslin cloth after collection to avoid moisture which may be fatal to flies.
3. Release the whiteflies in the cage through side opening after an infected leaf is inserted below the cardboard and leave them there for 1-4 h for acquisition access.
4. After acquisition access period, remove the infected leaf from cage by slight shaking so that the whiteflies remain within the cage.
5. Insert the leaf of the test plant in the cage similar to the infected leaf. Leave the vector overnight for transmission feeding.
6. Change the test plant after every 24 h, if required for about a week.
7. Spray all the experimental plants by insecticide and maintain in glasshouse for recording symptoms.

BEETLE TRANSMISSION

About 45 different viruses are known to be transmitted by 74 species of beetles. Beetles have biting mouthparts (leafhoppers and aphids have sucking mouth parts). Mainly comoviruses, bromoviruses, tymoviruses and southern bean mosaic virus are transmitted by beetles.

The acquisition feeding period in beetles may be from few hours to 24 h. Normally increased acquisition access period may increase transmission and the virus may be retained for longer period. There is no latent or incubation period in beetles and they can transmit the virus immediately after acquisition access period. Retention of the virus by beetles fall into two categories: some beetles remain viruliferous for one to two days and others 7-21 days. It is not clear whether beetle transmission is just mechanical or there is any biological relationship. The following protocol may be followed.

1. Collect the beetles in a test tube.
2. Release them to a Petri plate having diseased leaf. Cover the Petri plate with muslin cloth after releasing the beetles.
3. Give acquisition feeding for 24 h.
4. Release 5-10 beetles to caged test plants for transmission feeding for 24 h.
5. Remove or kill the beetles by insecticide and keep the test plants in the glasshouse for observation.

MEALY BUG TRANSMISSION

Nineteen species of mealybug are known to transmit about 7 viruses. *Planococcus njalensis* and *P. citri* are known to be the vector of most viruses. Cacao swollen shoot and citrus yellow mosaic badnavirus are good example of mealybug transmitted viruses. These insects are not mobile and hence are not efficient vector. They crawl from plant to plant but rarely.

1. Collect the mealy bugs in a Petri plate. Give fasting of 24 h.
2. Give 48-72 h acquisition feeding on infected plant.
3. Give transmission feeding of 24 h to test plants (release 10-15 insect/plant).
4. Spray the experimental plants with insecticide and put them in a glasshouse.
5. Mealy bug transmitted viruses require longer period for symptom expression. Hence, observe the inoculated plants for longer periods.

TRANSMISSION BY THRIPS

Tomato spotted wilt virus is a good example of thrips transmitted viruses. Four species of thrips transmit viruses in a persistent manner. *Thrips tabaci* is the major vector of these viruses.

The acquisition feeding period is 1-4 days. After acquisition, the insects take a period of 4-16 days to become viruliferous and then the insects can transmit the virus till they are alive.

PSYLLID TRANSMISSION

Psyllids (*Diaphorina* and *Trioza* species) are the vectors of some phytoplasmas (pear decline) and bacteria (Citrus greening bacteria). For transmission by psyllids, the following protocol may be followed.

1. Maintain healthy psyllids in wooden cages (described earlier) in the insectory on *Citrus* species or on curry leaf plant (*Murraya* sp.).
2. Collect the insect from the cage with the help of a catcher (described earlier).
3. Release the psyllids on an infected plant in a separate cage for acquisition access.
4. Give acquisition access of 24-48 h.
5. Shift the psyllids to healthy test plants in a separate cage (10 insect/plant)
6. Give 72-96 h transmission feeding.
7. Spray the plants with insecticide after transmission feeding and maintain in the glasshouse. Observe for symptom development.

Psyllids are the vector of citrus greening disease (a bacterial disease) in Asia and South Africa. In Asia, the vector is *D. citri* and in South Africa, *Trioza erytrae.* Psyllids also transmit pear decline phytoplasma disease.

Other insects like leafminer has also been reported the vector of tobacco mosaic virus but the mechanism of transmission is not known. The transmission may be solely mechanical. Beet leaf curl virus is reported to be transmitted by a lace bug in a circulative and propagative way.

MITE TRANSMISSION

Two types of mites are commonly invade plant species, the spider mite and the eriophyid mite. Eriophyid mites are the vectors of several plant viruses like pigeonpea sterility mosaic, peach mosaic, wheat streak mosaic, citrus leprosies and fig mosaic etc.

Eriophyid mites are elongated and tiny about 0.2 mm long. They can not be seen with naked eyes. They are found hidden between leaf hairs and buds. The virus transmission by mites is circulative or persistent type. Transmission studies by mites are comparatively difficult as they are microscopic. However, following protocol may be followed.

1. Collect leaves from a suspected virus infected plant.
2. Observe the leaves under a binocular for the presence of mites.
3. If mites are found, pick them up with the help of a specially made catcher. A very thin wire is fixed on a handle of a zero number painting brush (remove hairs from handle) with the help of lac.
4. Lift the mites one by one with the help of catcher under binocular and put on the leaf of healthy test plants for transmission studies or the mites may be collected on half cut citrus fruit for multiplication.

5. The offsprings from this collection can be used to determine the acquisition access period, incubation period and transmission feeding period.

TRANSMISSION BY SOIL-BORNE VECTORS

Soil inhabiting fungi like *Polymyxa* and *Spongospora* species and nematode species are known to be the vector of plant viruses. Tobacco stunt, soil-borne wheat mosaic, and potato mop top etc. are the example of fungal transmitted viruses.During transmission process, the virus particles either remain attached to the surface of the fungus or may be carried internally by the fungus. To carry out fungal transmission, it is important to learn the life cycle of the concerned fungus.

Nematodes play a major role in transmitting important viruses of two distinct groups, the tobraviruses (earlier known as netuviruses) *e.g*. Tobacco rattle virus and nepoviruses *e.g*. grapevine fan leaf and tomato ring spot viruses. Tobraviruses are transmitted by two genera of nematode, *Trichodorus* and *Paratrichodorus* and nepoviruses are transmitted by *Longidorus* and *Xiphinema* species.

Tobraviruses have an acquisition access period of about 15 min to 24 h. It is believed that the virus particles are attached to the mouthparts of the nematodes and as soon as they feed on healthy plants the virus is transmitted. The transmission is non-persistent type.

The vector, *Xiphinema* species feed on roots and form small galls over there. The vectors of nepoviruses can retain the virus for several days to several months. Nematode transmitted viruses are transmitted through seed and are also transmitted mechanically. Several nepoviruses have been reported to be pollen transmitted.

Following protocol is suggested for nematode transmission.

1. Collect the soil from the feeder root zone of the infected plant.
2. Bring the soil to the lab and give sieve washing to separate the nematodes.
3. Examine the concentration of nematodes under light microscope and get them identified by a nematode taxonomist.
4. Separate out virus transmitting genera and multiply each genus separately on healthy plants in laboratory for use in transmission studies.
5. Mixed the nematodes in soil, fill up pots with this soil and put young test plants in these pots.
6. Uproot the individual plant from the pots at various intervals (one after 24 h, second 48 h and so on), wash the roots thoroughly and replant in fresh pots having sterilized soil.

7. Alternatively, the rooted test plants are put in a water solution containing nematodes in a test tube. In this process also the nematodes can be given variable acquisition access and transmission feeding periods.
8. Maintain the test plants in the glasshouse and observe for the development of symptoms.

Care should be taken that water and fertilizers are clean and free from any nematode.

CHAPTER 4

HOST RANGE AND PROPERTIES OF THE VIRUS IN CRUDE EXTRACT *IN VITRO*

Once the methods of virus transmission are established, then we can study the reaction of the virus on different plant species by inoculating them with the best available method of transmission. This is referred as the host range of the virus. It is important during these studies that both local lesion and systemic hosts are identified.

To study the host range of the virus, seeds of plant species especially the cultivars of the original virus hosts are collected. However, common hosts of viruses like tobacco, cucumbers, leguminous plant species, *Chenopodium* species (for local lesions) etc. should also be collected from reliable sources. These seeds are grown in the sterilized soil in the glasshouse. Young seedlings of 3-4 weeks old should be used for inoculation. The methods of inoculation have already mentioned earlier. The symptoms developed in each plant species should be recorded for at least 4-6 weeks. The plant species which did not show any reaction should be tested (back inoculation) on a susceptible plant species (on local lesion host, if available) for any latent infection. Back inoculation is preferred especially when symptoms are confusing. A suitable healthy control should be maintained for each experiment.

The properties of the virus in crude extract can be undertaken with viruses which are mechanically transmitted and have good assay host(s). Following studies can be made in the leaf extract of virus infected plants.

THERMAL INACTIVATION POINT (TIP)

1. Grind infected leaves in appropriate buffer (normally phosphate buffer of 0.01 molarity and pH 7.6 is used as 1 g leaf to 1.5 ml buffer)
2. After grinding the leaves in sterilized pestle and mortar, filter the homogenate by muslin cloth.
3. Pipette out 0.5 ml sap into thin walled glass tubes.
4. Heat the individual sample at a particular temperature between 40-80°C at an interval of 5°C in a waterbath. The waterbath is put at required temperature and then sample tube is kept for 10 min. Treatment should start from higher temperature to lower as it will help to control the temperature easily.
5. After 10 min exposure to heat, the tube is removed from waterbath and kept into the cold water.
6. The sample is then inoculated to the assay host.
7. If a assay host is a local lesion host, results are recorded by taking the mean average of local lesions on at least six leaves of each treatment otherwise inoculate 3-5 systemic host plants for each treatment.

DILUTION END-POINT (DEP)

1. Extract the sap as mentioned in TIP and dilute it ten fold using distilled water or buffer. Each dilution is made by taking sap from a separate pipette. The dilutions are made as given below
   a. 1 ml sap + 9 ml distilled water (DW) = 10^{-1}
   b. 1 ml of 10^{-1} + 9 ml DW = 10^{-2}
   c. 1 ml of 10^{-2} + 9 ml DW = 10^{-3}

This is the method of making dilutions hence make more dilutions (up to 10^{10}. by this method.

2. Inoculate each dilution on assay host and record the local lesions with each dilution as has been described with TIP.
3. DEP is determined between a dilution which developed few lesions upon inoculation and a dilution at which no lesions were developed. For example if 2-3 lesions are developed at a dilution of 10^{-7} and no lesion at 10^{-8} then DEP would be between 10^{-7} and 10^{-8}.

LONGEVITY *IN VITRO* (LIV)

1. Extract the sap as described above and divide it in to a number of 0.5 ml aliquots in glass tubes. Each aliquot should have two tubes. One tube must be kept at room temperature and other in refrigerator (4°C).
2. Assay individual aliquot after every 24 h (may be up to 15-20 days)
3. Record the results of each sample maintained at room temperature and at 4°C separately.

The results are recorded that the TIP of the virus is between these temperatures (*e.g.* Between 55-60°C), DEP between this dilution to that dilution (*e.g.* 10^{-7} to 10^{-8}), and LIV between these days to those days (*e.g.* between 10 and 11 days).

CHAPTER 5

VIRUS ISOLATION AND PURIFICATION

PROPAGATION OF VIRUS

Before the virus is isolated it is important to multiply it in a suitable host. The host plant must be devoid of any inhibitors, should have higher concentration of the virus and can be grown easily. Plants like tobacco, cucumber, leguminous and Solanaceous plants are good virus multiplication hosts of many viruses.

To find out a local lesion assay host like *Chenopodium* species is important to determine the time of tissue harvest. The number of local lesions may indicate the concentration of virus in the leaf extract (inoculum). Harvesting is usually done 10-15 days after inoculation of an herbaceous host but this time may be longer for perennial hosts especially fruit trees. Usually leaves of inoculated plants are used but sometimes roots or flower petals are used when the leaves have higher concentration of inhibitors.

ISOLATION AND PURIFICATION

In the purification process virus particles are separated from host tissue and get concentrated *in vitro*. It is important that purified virus preparations should have more number of infective particles and devoid of host constituents. However, it is slightly difficult but one must try to obtain a preparation of maximum purity.

The first stage of purification is to harvest young symptomatic leaves and to crush them for getting an extract with virus particles. The crushing is preferred in chilled mortar and pestle or in liquid nitrogen. The extract is filtered in double layer of muslin cloth to remove cell debris. The filtered extract is given low speed centrifugation to remove heavier constituents of cells like chloroplast, mitochondria, starch grains which will settle down during low speed centrifugation (10000 rpm for 8-10 minutes) and can be discarded. The extract can also be clarified by heating the extract for 10 min at 50-60°C which will help to coagulate host proteins which can be removed by filtering through cheese cloth or by low speed centrifugation.

The most troublesome host materials to be removed from the extract are host proteins, ribosomes and microsomes as they resemble virus particles. After clarification of virus particles, the virus is concentrated in to a small volume which is referred as "partially purified preparation". This preparation still has host constituents and can be used for some diagnostic studies but needs further purification for removal of the host constituents.

The virus concentration is greatly increased if the extract is made in presence of a suitable buffer. The choice of buffer to be used for extraction will depend on particular virus. Therefore, the efficacy of buffers should be investigated experimentally by assaying extract in each buffer on a local lesion host. An example of turnip mosaic virus is shown in Table 2.

Table 2. Effect of different buffers on infectivity and stability of turnip mosaic virus

Buffer at pH 8.0	**Average no. of local lesions per leaf of *Chenopodium amaranticolor* after different periods of storage in hours at 22-25°C**			
	120	**144**	**168**	**192**
Kirkpatrick and Linder phosphate	3	2	00	00
Phosphate	24	19	3	00
Sodium citrate	5	3	00	00
Phosphate ascorbic acid	11	4	1	00
Citric acid	9	1	00	00
Sodium borate	3	1	00	00
Tris-HCl	23	17	2	00
Control	14	9	1	00

Note: The preparation of various buffer is given in Appendix-I

In the above experiment the sap extracted in phosphate buffer produced more lesions on leaves of *C. amaranticolor* and for longer period of storage. This is how a buffer is selected for extraction.

Similarly, the suitable molarity and pH of the buffer should be indentified. A buffer with suitable pH and molarity should be identified before the process of extraction is started. Molarity of 0.1 to 0.5 is normally used but the exact needs to be determined by experimentation. For every virus there is a pH at which the particles have no net charge (known as isoelectric point) and at this pH the particles may precipitate. Therefore, this pH must be avoided.

Several viruses are oxidized during extraction. To prevent this, reducing agents like thioglycollic acid, sodium sulphite or mercaptoethanol are normally added to the extraction buffer at low concentrations (around 0.1%). Chelating agents such as 1% EDTA (Ethylene-diamine tetra-acetate) of 0.05 M (pH 7.5) are also added to remove host ribosomes or polyribosomes. Other chelating agent is bentonite clay which can absorb ribosomes, chloroplast and ribonucleases. Pigments can be removed by addition of charcoal. Addition of 0.5-1 M urea or detergents like Triton X-100 or tween-80 help in release of virus particles from host constituents and reduce aggregation of virus particles.

CLARIFICATION

The organic solvents like ethanol, butanol, chloroform ether, and carbon tetrachloride are added to the aqueous phase obtained from low speed centrifugation to coagulate the larger host constituents without affecting virus particles in the extract. The concentration and suitability of each solvent is determined by experiments. For example, 6% cold n-butanol is suitable for many viruses but not for all. The higher concentration on n-butanol may harm the virus particles. Stirring after adding the solvent for about 30 min will help to improve the clarification. The host proteins are precipitated by adding the solvent which can be removed by low speed centrifugation (10000 rpm for 5 min).

CONCENTRATION OF THE VIRUS

The aqueous phase obtained after low speed centrifugation is used for clarification, which is again centrifuged in an angle rotor in an ultracentrifuge (about 70000 to 90000 x g to 120 min). In this process, virus particles are sedimented by forming a pellet at the bottom of centrifuge tubes. The aqueous phase normally referred as supernatant is discarded and virus pellet is suspended in a small volume (1-2 ml) of diluted buffer and allowed to stand for a few hours so that the pellet is slowly dissolved in buffer. This solution is given low speed centrifugation (5000 g for 10 min) and the supernatant is taken. This supernatant is diluted in larger quantity of buffer to fill the centrifuge tubes and given high speed centrifugation followed by low speed centrifugation as mentioned above. This method is called as *differential centrifugation.*

The virus in the supernatant after low speed can be further precipitated either by 30% PEG (polyethylene glycol) or by salt precipitation (a concentrated solution of 1/3 saturation of ammonium sulfate).

FINAL PURIFICATION

For final and highly purified virus purification, density gradient centrifugation technique is most commonly used. This is done in swing-out bucket rotor. The gradient of sucrose ranging from 10-40% is made. Sometimes, salts like cesium chloride or cesium sulfate gradients are also used for final purification where the virus particles are separated according to their buoyant densities. The preparation of density gradient is given in the method of purification of cucumber mosaic virus on following pages. Gradients are prepared by layering solution of decreasing density on top of one another and allow them to diffuse into each other overnight. The virus preparation (0.5-1.0 ml) obtained earlier is put on the top of the gradient and centrifuged in a swing-out bucket rotor (50000-70000 x g). The virus particles are separated from contaminants through this procedure as they will sediment according to their sedimentation coefficient.

After centrifugation virus layers can be observed in a different beam of light within the centrifuge tube (Fig. 43). The virus can be collected manually with a hypodermic syringe or in a density gradient fractionator. The sucrose or salt can be removed by dialysis and virus can be re-pelleted by high speed centrifugation or by other methods like gel-chromatography, electrophoresis or antisera produced against host proteins (to absorb host proteins from virus preparation).

An example of a protocol used for purification of a filamentous virus [Indian citrus ring spot virus (ICRSV)] is given below.

1. 50 g young symptomatic leaves from ICRSV inoculated French bean plants were harvested after 2-3 weeks of inoculation. Washed them and cleaned with tissue paper.
2. The leaves were homogenized in cold and sterilized pestle and mortar with 10 volumes of extraction buffer (0.05 M phosphate buffer containing 0.005 M DIECA, 0.01 M EDTA and 0.02 M sodium sulphite).
3. Filter the extract through double layer of muslin cloth.
4. Added 10% chloroform of the total volume of filtered extract, stir it for 5 min at room temperature.
5. Centrifuge the chloroform and extract mixture at 12000 rpm for 10 min in a SRC-5 rotor at 4°C in a low speed centrifuge.
6. The pellet is discarded and the aqueous phase is centrifuged at 38000 rpm for 3 h in a Beckman 55.2 Ti rotor at 4°C.

7. Suspended the pellet in 2 ml of extraction buffer.
8. Centrifuged at 1000 rpm for 2 min.
9. Retained the supernatant. Resuspended pellet in 1 ml of extraction buffer, centrifuged at 1000 rpm for 2 min (to isolate particles, if settled with pellet). Mixed both the supernatants.
10. Layered the supernatant (1.5 ml) on cesium sulfate gradient (10-40 % prepared in extraction buffer).
11. Centrifuged at 41000 rpm for 2.5 h in a Beckman SW 41 rotor (swing-out bucket rotor).
12. Collected the single virus band through ISCO density gradient fractionator and diluted in extraction buffer.
13. Centrifuged at 70000 rpm for 30 min in a Beckman R-70 rotor.
14. Resuspended the pellet in 200 µl of 0.01 M phosphate buffer, pH 7.2 and used as final preparation for serology and electron microscopy.

Some viruses especially of PVY group can be concentrated effectively when in the supernatant obtained after first low speed centrifugation, PEG-600 and NaCl are added to a final concentration of 4.0 and 1.75% (W/V) respectively. The protocol for PVY can be followed as:

1. The PEG and NaCl mixture after stirring for 30 min was allowed to stand at 4°C for one hour.
2. Centrifuge at 8000 g for 12 min.
3. The precipitate was resuspended in 1/5th of the original volume of 0.5 M borate buffer, pH 8.0 with overnight slow agitation at 4°C.
4. Next day, the material was centrifuged at 5000 g for 10 min and pellets were discarded.
5. The supernatant was layered on 30% sucrose pad (W/V) made in borate buffer and pelletised at 80,000 g for 2h in Beckman 55.2 Ti rotor.
6. Pellets were resuspended in 0.05 borate buffer containing 5 mM EDTA.
7. Differential centrifugation steps are repeated, if required.
8. Density gradient centrifugation (sucrose density gradient) is done as mentioned earlier and virus zone is removed.
9. The fractions collected in the fractionator were dialysed overnight at 4°C with three changes with 0.05 M borate buffer, pH 8.0 containing 5 m M EDTA.
10. The virus was then pelletised at 80,000 g for 2 h in Beckman 55.2 Ti rotor.
11. Pellets were resuspended in small volume of suspension buffer.

The purification of filamentous viruses has been shown above. The purification of an icosahedral (spherical) virus (cucumber mosaic virus) is given below.

1. Homogenize 200 g of infected symptomatic cucumber leaves in cold 400 ml of 0.5 M sodium citrate buffer at pH 6.5 containing 0.4 ml thioglycollic acid and 400 ml of chloroform.
2. Centrifuge the homogenate at low speed (500 x g) in a bench centrifuge for 15 min.
3. Add 10% (W/V) PEG-600 to the supernatant and put on a shaker for dissolving and then store for 30 min.
4. Centrifuge at 8000 x g for 20 min.
5. Discard supernatant, dissolve pellet in 0.05 M citrate buffer (pH 7.0) containing 2% triton X-100.
6. Put 3.5 ml buffer with virus particles to each centrifuge tube (3.5 ml capacity) and leave them overnight.
7. Centrifuge at 15000 x g for 20 min, retain supernatant and discard pellet.
8. Centrifuge the supernatant at 75000 x g for 30 min in an ultracentrifuge. Discard supernatant and dissolve the pellet in 1 ml of 0.05 M citrate buffer (pH 7.0), leave overnight.
9. Centrifuge the resuspended pellet at 5000 x g for 10 min.
10. Discard pellet, test the infectivity of supernatant (partially purified preparation) and examine the concentration of the virus in electron microscope, if available.
11. Further purification can be done by passing through sucrose gradient (10-40%) which can be prepared as
12. Dissolve sucrose in distilled water or in a suitable buffer (0.01-0.1 M).

    5 g sucrose in 50 ml DW = 10%

    10 g sucrose in 50 ml DW = 20%

    15 g sucrose in 50 ml DW = 30%

    20 g sucrose in 50 ml DW = 40%
13. Add sucrose solution in 9 cm x 2.5 cm cellulose nitrate centrifuge tubes with the help of a pipette.
14. First add 7 ml of 10% sucrose solution in such a way that the tip of pipette remains in contact of the bottom of centrifuge tube. Then in a similar fashion 7 ml of 20% solution and successively 30 and 40% solutions so that 40% remains at the bottom and 10% at the top.

15. Store the tubes in a refrigerator overnight so that the solutions are diffused and a gradient is formed.
16. Add 0.5 ml of partially purified virus preparation over the gradient along the side of the tube.
17. Centrifuge the gradient in a swing-out bucket rotor at 41000 for 60-90 min. Care should be taken that the tubes fixed in rotor are perfectly balanced against each other.
18. Follow the procedure as mentioned earlier to get highly purified preparations.

By modification in buffers, their molarity and pH, organic solvents, speeds of centrifuge and use of chelating agents etc. can be used to purify majority of viruses. The purified viruses can be used to determine their morphology and size in the electron microscope and production of antisera. Various physico-chemical properties of the virus can also be studied with purified preparations.

PHYSICO-CHEMICAL PROPERTIES OF PURIFIED VIRUS PREPARATION

The physical properties help in recognizing known viruses and identifying new ones. Among them commonly determined properties are to determine the size and morphology by electron microscopy, and absorption of ultraviolet radiation (A) used to estimate virus concentration. The optical density is usually measured by comparing the intensity of the light transmitted by the virus suspension with that by the fluid used. The absorption spectra maximum for proteins at 280 nm (A280) and for nucleic acids it is at 260 nm (A260). The ratio of A_{260}/A_{280} can therefore be used to estimate the content of nucleic acid in a virus.

DETERMINATION OF ULTRA-VIOLET ABSORPTION PROFILE

20 µl purified virus preparation is diluted 25 times with 0.05 M phosphate buffer at pH 7.2 to make the volume 500 µl (The buffer suitable for a particular virus is used for dilution). The quartz cuvette is filled with phosphate buffer and calibrated to zero absorbance in a spectrophotometer. After calibration, the cuvette is taken out, emptied and filled with 500 µl of virus preparation. UV absorption is taken at 260 nm and 280 nm in spectrophotometer. For example, results of four purified preparations of Indian citrus ring spot virus are given in Table 3.

These results indicated that the preparations used in this experiment were nucleoproteins.

Table 3. UV absorption of virus preparations at 260 and 280nm

Preparation Number	UV absorption		A_{260}/A_{280} ratio
	260 nm	280 nm	
1	0.838	0.750	1.12
2	0.760	0.694	1.10
3	1.450	1.370	1.06
4	1.570	1.358	1.16

CHAPTER 6

ELECTRON MICROSCOPY

The Electron Microscope

The transmission electron microscope (TEM) (Fig. 44) has been an indispensable tool in biological, medical as well as material research. Operating on the basic principles of the light microscope, the TEM takes advantage of the much shorter wavelengths of electron beams providing resolving powers on the order of 0.2 nm, rather than 200 nm for the light microscope.

In the light microscope a light beam from a special lamp is focused through the specimen by the lens. The light reaches the objective lens, which magnify the image of specimen. The eyepiece further magnifies and projects the image to the eye or a photographic film. The smallest objects seen with the glass lens system using visible light are bacteria. To obtain much finer detail than light microscope, a beam of shorter wavelength must be used. X-rays are much shorter but cannot be focused, as they have no charge.

To overcome this problem electron beam is used as a source of illumination. Extremely small negatively charged electrons given off by a heated tungsten filament can be accelerated by high voltage to produce a coherent electron beam and can be focused by an electric field.

The electric field of a doughnut-shaped electromagnet surrounding the electron beam acts just like the glass lens, which focuses the light beam on the specimen. A similar electromagnet is used as the objective lens to form a highly magnified image of the specimen; and one or two electromagnetic lenses further magnify and project the image onto a fluorescent viewing screen or photographic plate or light sensitive sensor such as a CCD (charge-coupled device) camera. The image detected by the CCD may be displayed in real time on a monitor or computer. Magnification is increased in the light microscope by inserting stronger objective or projector lenses. In the EM, strengthening the projector lenses increases magnification by increasing the current through electromagnets.

Since the electrons are very much smaller than air molecules so there must be high vacuum inside the column to allow a free path for the electrons from filament to the viewing screen. The specimen should also be very thin to allow the electrons to pass through to form an image. Special methods have been developed to observe very small particles and solid surfaces. Tissues are fixed and resin blocks are prepared to cut the sections by ultra-microtome using glass or diamond knife. Both particles and ultra-thin sections are held in the electron beam on carbon-coated grids. These grids are thin enough to be transparent to the electrons. The grid is held in a movable holder for observation in the EM.

Following are the common methods used for virus detection in electron microscopy:

LEAF DIP METHOD

It is very quick and reliable test for the presence or absence of virus(es) in infected plant tissue. This method is very useful for the detection of those viruses that remain in very high concentration in leaf tissues, for example Tobamoviruses, Potyviruses etc.

Method:

1. Cut out a 2 x 3 mm piece of infected tissue with a blade or peel off an epidermal strip from the underside of the infected leaf.
2. Using the forceps, dip the tissue sample into the stain drop (2% uranyl acetate, pH-4.5) for one or two seconds, taking care not to damage the support film on the grid.
3. The excess of stain can be removed by very lightly touching the edge of the stain drop with pointed strip of filter paper.
4. The virus particles will be visible when viewed the sample in electron microscope.

LEAF SQUASH METHOD

This method allows large pieces of leaf to be used, but leaves large amount of cell debris on the grid.

Protocol

- Grind an infected leaf of about 2-3 mm diameter over a clean glass slide in phosphate buffer, 0.07 M, pH-6.5
- Place 10 µl of homogenate on para film or waxed slide in a wet Petri plate.
- Place 400-mesh carbon coated copper grids (film side downward) on the surface of the droplet, ensuring that the grid surface is wet. Allow it for 2-5 min.
- Pick up the grid by its edge with fine forceps and wash the grid with 10-15 drops of double distilled water (DDW) to remove the sap.
- The grids are stained with 2% aqueous uranyl acetate (UA). The excess stain is immediately drained off by using Whatman filter paper.
- The grids are observed under EM to see the virus particles.

The virus particles have different morphology which can be determined by electron microscopy (Fig. 45).

Measurement of virus particles

It is important to determine the modal length of the virus particles. It is done by taking electron micrographs of virus particles at a magnification on the plate film or on 35 mm film. The micrographs are magnified according to the requirements. The measurement of length of at least 100 particles in mm is taken from the negative and the actual length of particles is calculated either by "ultra size calculator" (Tedpella Inc. P.O. Box 510, Tustin, California, USA) or by using the following formula:

$$\text{Size of virus particles in nm} = \frac{\text{Particles size in mm}}{\text{Magnification}} \times 1{,}000{,}000$$

Note: The measurements of the particles should be taken from micrographs of leaf dip preparations. However, if it is difficult to take from leaf dip then count 100 particles from micrograph of purified preparation.

ELECTRON MICROSCOPY OF ULTRA THIN SECTIONS

Method for processing tissue for ultramicrotomy

The material to be fixed (Leaf, stem or root) is thoroughly washed with distilled water. Cut small portion of the tissue (1.5 x 3 mm) by using sharp razor blade. Cut tissues are transferred to vial containing 2.5% glutaraldehyde (GA) using fine forceps. Vacuum infiltration of tissues is done till tissues sink down. Vials are kept at 4°C for 24 h (for harder tissues duration is more).

Tissues are thoroughly washed with phosphate buffer 0.1 M, pH 7.0. Secondary fixation is done by immersing tissue in 0.2% osmium tetraoxide (OSO_4) in vials and kept at 4°C overnight (O/N).

Tissues are dehydrated at room temperature as given below.

a. 25% acetone – 2 changes for 15 min each.
b. 50% acetone – 2 changes for 15 min each.
c. 70% acetone- Tissue are kept in mixture of ½ vol. 70% acetone + ½ vol. 70% acetone saturated with uranyl acetate at 4°C O/N.
d. Final 2 changes are given in 100% acetone for 15 min. each.
e. Put tissue in propylene oxide to remove acetone for 5 min.

Infiltration:

Step 1: Immerse tissue in a mixture of 1 part of resin + 3 parts propylene oxide at 40°C for 1 h and kept on shaker.

Step 2: Transfer to a mixture of part of resin + 1 part propylene oxide at 40°C for 1 hr on shaker.

Tissues are then transferred to a mixture of 3 parts of resin + 1 part of propylene oxide using conditions as before.

Tissues are transferred to pure resin and 3 changes are given after 1 h each at 40°C using shaker.

At each step new vial is used and transfer is done without injuring tissue.

Pre-embedding

Fix the washed tissue of 1 x 1 mm size in 1% GA diluted in 0.1 M PO_4 buffer (pH 7) for 1 h, vacuum infiltration for 10 min. and keep in GA for 24 h.

Wash the tissue with PO_4 buffer for 15 min.

Incubate the tissue with 1% antibody diluted in PO_4 buffer containing 1% BSA & 0.1% Tween 20 (PB-BT) at RT for 2-4 h.

Remove liquid with Pasteur pipette, rinse with PB-BT for 20 min. Incubate with Goat anti-rabbit IgG-10 nm gold diluted 1/50 in PB-BT at RT for 24 h.

Rinse in PB, Fix in 0.2% osmium tetraoxide (OSO_4) for 1 h complete the embedding in epoxy resin as described earlier. Cut sections and stain in uranyl acetate (UA) and lead citrate.

Embedding

Embedding of tissue is done in pure resin using flat silicon moulds. Moulds are just filled with resin and tissues are placed at one end either in a transverse or longitudinal orientation. Resin is then polymerized at 60°C overnight. Moulds are taken out and allowed to cool down on bench. After cooling blocks are taken out of moulds. They are ready for sectioning after 24 h.

Post-embedding

Prepare blocks as described above.

Cut sections as described in 2.3 of 90 nm thickness and collect on nickel grids.

If OSO_4 is used, treat sections for 5-6 min, with a saturated solution of sodium metaperiodate ($NaIO_4$); wash in distilled water for 5 min.

Treat sections with primary antibody diluted in phosphate buffer containing 0.1-1% BSA, 0.1% Tween-20 and 0.05% sodium azide. This buffer is commonly called as PB-BT. Around 10-100 ug/ml IgG is optimal conc. for primary antibody.

Rinse in PB-BT, incubate with gold labeled antibody (commercially available) 30-60 min. (usually gold of 10 nm size is used & it is diluted 10-50 times in PB-BT.)

Rinse in water. Stain first in 5% UA and then in 5% lead citrate.

ULTRATHIN SECTIONING OF SPECIMEN FOR ELECTRON MICROSCOPY

Trimming of blocks

Trimming of the blocks is done with the help of a sharp razor blade to make them to a pyramid shape.

After face of the block is made, thick sections (0.5 to 2 um thick) are cut by razor blade. They are taken on glass slide and seen under optical microscope. This enables to see

a. State of embedded specimen
b. The fixation
c. Selection of area for ultrathin sectioning.
d. The size and position of cutting face for final trimming

After scanning sections under optical microscope, the area to be examined under TEM is selected and blocks are further trimmed, if required.

Alignment

After final trimming, fresh face of the block is made. The block face should be smooth and shiny. Now fix glass or diamond knife in the specified place. Adjust its height and clearance angle according to specification. Lock the orientation head containing block in the well in the ultra microtome. Observe the knife and block through microscope. The reflection of the knife edge in the block face should be completely parallel and horizontal with the real knife edge.

Cutting

Using the instruments as per ultra microtome make & model the cutting speed is set. The sections are cut. Various colours of the sections in the knife trough will be seen.

Gray – 60 nm thick – The sections are useful for high resolution work.

Silver – 60-90 nm thick – The sections are ideal for seeing ultra structures under TEM

Gold – 90-150 nm thick – The sections are good for seeing under low magnification for pathogens like phytoplasmas.

Purple, blue, green and yellow are thick sections and are usually not used.

The sections are manipulated in position using mounted eyelash and then transferred to grid by either carefully scooping them from below or touching from above. If grids are of large mesh size (100 or 200) then a coating of plastic film is essential to render support to the sections. Film coating is done using 0.6% formvar (Polyvinyl formaldehyde) in ethylene dichloride

The grid is drained of excess water using Whatman paper strip and viewed in electron microscope. The virions may be visible in tissues of the section. The technique is useful to know the location of the virus in tissues. The test is used both for viruses (Fig. 46) phytoplasmas and fastidious bacteria (citrus greening bacterium (Fig. 47).

PREPARATIION OF GLASS KNIVES

I. To make squares from long strips:

A. Setting

1. Retract rear glass holder
2. Set white guide at 90°.
3. Set score selector on symbol.

4. Turn breaking knob (on front panel) to counter clockwise stop.
5. Set damping device at "0".

B. Operation

1. Place glass strip, score side down, against guide plant. If strip is short, place against stop. If strip is full length, break in half.
2. Hold glass strip against plate. Lower damping head. As soon as head contracts glass, transfer hand pressure from glass strip to head. Do not over-tighten head.
3. Insert fork.
4. Score rapidly with shaft.
5. Turn breaking knob clockwise immediately and rapidly. Return knob to counterclockwise stop after break.
6. Raise clamping head with both hands; one on lever and one supporting head.
7. Push scoring shaft back.

II. To make knives from squares.

1 Settings

1. Set score selector.
2. Set front glass holder, rear glass holder, and dials and to set as recommended on machine.
3. Insert square between white bumpers on holder.
4. Insert fork.

B. Operation

1. Score as in Step I.B.4.
2. Turn damping lever until damper just touches glass. Do not exceed 45°.
3. Break as in I.B.5,6,7.
4. Return damping lever to "0"
5. Retract back glass holder mechanism and remove knives.
6. Inspect knives.
7. Inspect knife edges critically with vertical illumination.

2 Staining Method

Uranyl Acetate

5% UA solution is made in double distilled water. It is stored in amber coloured bottle.

Lead citrate

Add half a pellet of NaOH to 10 ml of DDW in tight screwed bottle. Shake well. Add 0.5 gm of lead citrate. Shake well, seal it with parafilm and store at 4°C.

10 ul of uranyl acetate is dropped on parafilm (cover with petridish) and grid is floated on this drop from the side of section for 10 min. Then sections are washed with 15-20 drops of DDW, Grid is drained off excess water.

10 ul drop of lead citrate solution is kept on the parafilm (covered with a petridish). 4-5 pellets of NaOH is kept inside the petridish. NaOH keeps the air free of CO_2. CO_2 reacts with the lead citrate to form electron dense globules of lead carbonate. Grid is floated on this Whatman paper strip. Sections are now can be screened under TEM.

3. Materials required in electron microscopy

3.1 *Chemicals*

1. **Sorensen's phosphate buffer – 0.1 M, pH 7.0**
   a. Sodium dihydrogen phosphate (NaH_2PO_4) Mol. Wt.-156.01. Dissolve 6.084 g in 200 ml of S.D.W. and make volume to 390 ml
   b. Disodium hydrogen phosphate (NaH_2PO_4) Mol. Wt – 178.00. Dissolve 10.85 gm in 500 ml of D.D.W. and make volume to 610 ml. Mix a & b together, adjust pH to 7.0.
2. **Glutaraldehyde:** It is commercially available as 25% or 70% stock solution. Take 1 ml of 25% GA and add 9 ml of PO_4 buffer pH 7.0. Make solution in amber coloured bottle and stored at 4°C. As its fumes are toxic, solution should be handled in fume-cupboard.
2. **Osmium Tetroxide:** It is available in ampoules containing 1 gm or 0.1 gm. Ampoule is scored using diamond cutter and ampule is transferred to big amber coloured bottle. It is pounded using glass rod and enough PO_4 buffer (0.1 M, pH 7.0 is added to make 0.2% osmium tetroxide solution. Bottle is sealed with parafilm & stored at 4°C. It is toxic & solution should be made under fume-cupboard.
3. **Acetone:** Analar Grade
4. **Propylene Oxide:** Analar Grade
5. **Resin:** Many types of resins are commercially available. Commonly used are Aradite CY 212 and Epoxy epon 812, which are highly viscous resins. Epoxy epon 812 is made by adding 4 components in plastic beaker and mixing them thoroughly for 10 min. As it is highly expensive and can be used only once, only required amount should be made fresh.
   a. Methyl Nadic Anhydrite (MNA) – 25 g. It is softener (Mol. Wt = 178.9)

b. TAAB epon 812 resin – 50 g (Epoxide equivalent = 150)

c. Dodecenyl succine Anhydride (DDSA) – 5 g. It is hardener (Mol. Wt. = 266.38).

d. 2,4,6 tri (Dimethyl amino methyle) Phenol DMP = 1.5 g. It is an accelerator.

The block can be made hard or soft by varying amounts of softener or hardener. The resin are proven carcinogen so disposable gloves should be used during handling.

6. Uranyl acetate – EM Grade
7. Lead citrate – EM grade
8. Bovine albumin serum
9. Tweem – 20
10. NaOH Pellets
11. Gold label – 10 nm size – Conjugated with anti rabbit IgG.

Glassware

It should be clean and autoclaved.

Glass vials, Pasteur pipette, Measuring cylinder, Beaker.

Miscellaneous

EM Grids – 300 mesh size, Whatman Paper, Parafilm, Fine foreceps, Camlin Brush, Razor Blade, Scalpel Handle, Aluminium foil, Disposable hand gloves.

3 Precautions

The chemicals used for fixing, embedding and staining are expensive and bio-hazardous. Proper care should be taken while handling them.

All the experiments should be carried out in dust free environment.

Temperature maintenance is important for proper functioning of TEM and ultramicrotome.

All the things required for first-aid should be handy.

CHAPTER 7

SEROLOGY

The phenomenon of immunity in an individual is known since many years that when an animal gets infected with a virus it rarely catches the same virus again. The immunizing agent (virus) is called *antigen* or *immunogen.* So the antigen in this case is a viral protein that stimulates the production of antibodies after injection into an animal. All the proteins are not good *immunogens* but many plant viruses are efficient *immunogens*. The antibodies are produced in the lymphatic tissue of animal. Such antibodies are specific and react only with protein (antigen) that stimulates their production and such antibodies are referred as *homologous* antibodies. The blood serum that contains antibodies is referred as *antiserum*.

The recognition proteins found in the serum that react specifically with the antigen that produced their formation are called antibodies. Antibodies belong to family of globular proteins called immunoglobulins. The terms antibody and immunoglobulin are used interchangeable. The antigenic determinants of antibodies show three levels of variability. The determinants are called isotypic, allotypic or idiotypic and classify antibody molecules into isotype, allotype and idiotype respectively. Isotypic determinants divide antibodies into five classes or isotypes called IgG, IgA, IgM, IgD and IgE.

The antibodies belong to a group of protein known as γ-globulins. These antibodies are called IgG which are Y-shaped and most commonly used by plant Pathologists for detection of plant pathogens. The reaction between antibodies and antigen paved the way for detection of plant viruses and these techniques are referred as serological techniques.

In a serological test, the antibodies react specifically with their homologous antigen by fitting together in a unique way (lock and key way). Antibody molecules have determinants or paratope sites on their surfaces which are complementary in shape, charge, and hydrophobicity to the epitopes of antigen surface. An antibody has two combining sites but the antigen molecule has numerous combining sites on its surface depending on its size.The antibody-antigen reaction is visible in the form of precipitate either in tube or in agar gel.

Antibodies are produced by injecting the purified virus (antigen) in the animal system. The animals used for this purpose are mice, rabbit, goat, sheep and horse. In the animal cells, the common sites for antibody production are lymphatic and plasma cells, spleen and bone –marrow. Following types of antibodies are known.

Polyclonal antibodies

Polyclonal antibodies represent the antibodies from multiple clones or β-lymphocytes, and therefore bind to a number of different epitopes. For antibody production purposes, these are obtained from the serum of an immunized animal, thereby providing a wide range of antibody types and epitope affinities (to molecules other than the intended antigen as well). For many applications, this diversity of antibodies provides an advantage by allowing the researcher to detect multiple epitopes sites on the protein of interest.

Monoclonal antibodies:

Monoclonal antibodies, by contrast, represent a single β-lymphocyte generating antibodies to one specific epitope. Two primary advantages arise with such an application. First by fusing the lymphocyte with an immortal myeloma cell line (both obtained from mice). One can create a hybridoma cell line capable of producing virtually unlimited quantities of a specific antibody. Second, the antibodies produced are identical and specific only to the epitope of interest.

Structure of an antibody

The antibodies are γ-globulins and the main one is IgG which is mostly used in plant virology (Fig. 48). It is composed of two long normally called as heavy polypeptide chains and two short polypeptide chain or known as light chains. These chains are bound by sulphydryl bonds (S-S). The variable regions are responsible for specific binding with matching antigenic determinants of the antigen. The two arms of y molecule can open or close at the hinge region to bridge antigenic sites. Each polypeptide chain has secondary or tertiary structures stabilized by S-S bond.

PRODUCTION OF POLYCLONAL ANTIBODIES

Following steps are used for the production of polyclonal antibodies in rabbit.

PREPARATION OF ANTIGEN

Highly purified virus preparation is mandatory for production of high quality antisera. If the virus is not properly purified then antibodies will also be developed against the impurities. Such antisera will interfere to diagnostic system and desired results can not be obtained. The purification procedures have been described in the chapter of virus purification.

IMMUNIZATION OF ANIMALS

Mostly rabbits are used for the production of antisera against plant viruses. The titre of antisera depends on the strain of rabbits. The most commonly used rabbit strain is called as *White albino rabbits*. The antigen may be injected either through ear vein (intravenous) or in to thigh muscle (intramuscular). If the antigen is highly purified then intravenous injections are given several times in small doses. The number of injections and amount and nature of immunogen to be injected determines the titre of antibodies. Intravenous injections are given at an interval of 2-3 days and intramuscular injections are given weekly. In total, 5-6 injections followed by a booster injection are given.

The antigen to be given intramuscularly is emulsified with equal volume of Freund's incomplete adjuvant. The adjuvant helps to slow release and uniform distribution of the virus to antibody production sites.

COLLECTION AND CLARIFICATION OF ANTISERUM

After 4-5 days of last injection, the rabbit is bled by giving a small cut to the marginal vein of the ear. About 20 ml blood is collected in a beaker and allowed to clot at room temperature. The serum is collected by decantation and clarified by low speed centrifugation (to remove RBC). The titre of the antiserum is checked. The blooding can be done 3-4 times with weekly intervals and grade the antisera as per their titre so that different grades are used effectively for different serodiagnostic tests.

If antiserum is produced by partially purified preparations then antibodies will also develop against host plant constituents and will give non-specific reaction in diagnostic tests. The commonly found constituents are host fraction-1 protein and phytoferritin. To overcome this problem antibodies are produced against the antigens of uninfected healthy plants. These antibodies are used to adsorb the antibodies formed in virus antiserum against plant constituents.

STORAGE OF ANTISERA

The antisera will remain workable for a long period if stored properly. The antisera may either be frozen or lyophilized in small vials (0.25-0.5 ml) and stored in vaccum or liquid nitrogen. To avoid crystallization of antibody molecules at low temperatures, the antiserum should be mixed with equal volume of glycerol. If the antiserum is stored in the refrigerator at 4°C for short periods then small quantity of sodium azide is added to check the bacterial growth.

PROTOCOL TO PRODUCE POLYCLONAL ANTIBODIES

1. Take 1 ml of purified virus preparation and emulsified with 1 ml of Freund's incomplete adjuvant.
2. Inject intravenously to the ear vein or intramuscularly in thigh muscles.
3. Give 6-8 injection at an interval of 4-6 days on alternate thigh.
4. After 8-10 days of last injection, bled the rabbit giving a small incision to ear vein. Collect about 20 ml of blood in a beaker. The blood may be collected 3-4 times with weekly interval so that enough quantity of antiserum is available.
5. Give booster injection after two weeks of last injection to get high titre antisera.
6. The blood in the beaker is covered and left for 24 h at room temperature to coagulate.
7. Remove the serum (antiserum) by decantation and give a low speed centrifugation (500 g for 5 min) to remove RBC. The supernatant is a clear antiserum.
8. Store the antiserum in small vials by adding glycerol 1:1 (V/V) or freeze dry (also see storage of antisera above).

Each β cell is programmed to make one specific antibody when a β cell encounters its triggering antigen (along with collaborating T cells and accessory cells), it gives rise to many large plasma cells. Every plasma cell is essentially a factory for producing antibody. Each of the plasma cell descended from a given β cell (which are all members of the same family or clones) manufactures millions of identical antibody molecules and pours them into the blood stream (Fig. 49). A given antibody matches an antigen much as a key matches a lock (Fig. 50).

The polyclonal antibodies are broad spectrum as they are produced against all the epitopes of virus proteins. However, it is difficult to detect strains of a virus by polyclonal antibodies and to develop polyclonal antibodies against phytoplasmas and some bacteria like greening bacterium are difficult as they are pleomorphic in nature. Because these organisms have no definite shapes, the purification is difficult. Therefore, monoclonal antibodies were developed which proved to be very useful for detection of these pathogens as described below.

MONOCLONAL ANTIBODIES

In 1975, Kohler and Milstein first fused *Lymphocytes* with *Myeloma* (tumor) cells to produce a cell line which was both immortal and a producer of specific (monoclonal) antibodies. The two scientists were awarded the Nobel prize for medicine in 1984 for the development of this *Hybridoma*.

Monoclonal antibodies are:

1. Antibodies of exceptional purity and specificity.
2. They are the components of the immune system
3. Able to recognize and bind to a specific antigen

Monoclonal antibodies are produced in mice. Spleen cells of mice, previously immunized with antigen (virus) are fused with mouse myeloma (tumor) cells. The fusion results in the production of hybrid cells which are known as hybridomas. The spleen cells of mice can not be grown in synthetic media and will die there but myeloma cells grow indefinitely in the medium. Therefore, a hybrid having two components, the spleen cell at the top (which produce antibodies) and the myeloma cells at the bottom (which facilitate the hybrid to grow in synthetic media indefinitely). Thus hybridomas are continuous source of antibody production. Hybridomas producing monoclonal antibodies (MAbs) of interest are selected through screening.

The hybridomas can be preserved by kryopreservation for unlimited periods. There are two major advantages of this novel technique (i) to produce MAbs against any antigenic determinant (ii) impure pathogen preparations can be used. Serodiagnosis of phytoplasmas could be possible only after the development of hybridoma technology. MAbs have been found very useful for the identification of virus strains and to study the structural and functional aspects of gene products.The protocol to develop monoclonal antibodies is shown in Fig. 51.

IMMUNIZATION OF MICE

The mice used for MAbs production is known as BALB/c strain of mice. The antigen is injected in to mice by the methods:

1. **Intraperitoneal:** The mice is injected at the lower part of the abdomen using 23 G needle. The care should be taken that the needle does not rupture the intestine.

2. **Subcutaneous and intramuscular:** Subcutaneous injection is given at the hide-limb flap around the neck. Hold the fold of loose skin and insert the needle at the base of the fold. The intramuscular injections are given to thigh muscles.
3. **Intravenous:** Antigen is injected into the caudal vein which is dilated with xylene.

Subcutaneous injection is usually given with 20-200 µg of antigen with Freud's incomplete adjuvant (50:50). After one month, booster dose is given with the same volume of antigen intaperitonially. Harvest spleen 3-5 days after booster injection by killing the animal.

Selection of myeloma cells

Use myeloma cells that have lost the ability to synthesize hypoxanthine guanine phosphoribosyl transferase (HGPRT). This enzyme enables cells to synthesize purines using an extracellular source of hypoxanthine as a precursor. Ordinarily, the absence of HGPRT is not a problem for the cells that have an alternate pathway that they can use to synthesize purines. However, when cells are exposed to aminopterin (a folic acid analog), they are unable to use this other pathway and are now fully dependent on HGPRT for survival.

Growth of Myeloma cell line

5-7 days before the fusion is planned, remove 1 vial of myeloma cells from the liquid nitrogen freezer (desired myeloma cells are preserved in vials in liquid nitrogen), thaw at 45°C and wash with 10 ml of RPMI-1640 medium.

Seed the cells in to 20% (fetal calf serum (FCS) and RPMI-1640 in a sterile Petri plate. When cells cover 50% of the surface of the Petri plate, transfer to 20 ml 10% FCSRPMI in 250 ml culture flask. The cells are then washed and 90% liquid is removed, then add fresh 10% FCS + RPMI.

FUSION PROTOCOL

1. Add 5 g PEG-400 to a beaker, heat gently (do not exceed temperature over 45°C) with stirring until PEG dissolves.
2. Cool at room temperature and add 1 ml dimethyl sulfoxide (DMSO).
3. Filter through 0.45 µ syringe filter into sterile tube. Place this fusogen at 37°C in a water bath.
4. Harvest myeloma cells and place them in one 50 ml conical tube and centrifuge approximately at 800 g for 10 min. Decant off supernatant.
5. Suspend pellet in 10 ml RPMI and then centrifuge as in step 4.

6. Dissect immunized mice and remove spleen (aseptically) and place it in sterile Petri plate with 10 ml of RPMI. The spleen may be put between two glass slides and the slides are rubbed with each other so that spleen cells are spread over slides. Harvest cells by Washing slides with 10 ml of RPMI.
7. Centrifuge harvested cells as earlier.
8. Mix spleen cells and myeloma cells together in round bottom 12 ml tube and centrifuge.
9. Immerse bottom of the tube with pellet into 37°C water bath.
10. Add 1 ml 50% PEG, 10% DMSO-RPMI over 1 min while stirring the pellet with the pipette tip.
11. Add 1 ml RPMI medium over 1 min slowly to dilute out the PEG solution
12. Add 2 ml RPMI over I min.
13. Add 6 ml RPMI over 3-5 min.
14. Spin down cells, and resuspend them in 5ml HAT media (20% FCS, RPMI supplemented with aminopterin, thymidine and hypoxanthine).
15. Sacrifice 6-week old 2-3 mice and remove the thymus. Release the thymocytes by the step 6 above. Mix thymocytes with the fused cell mixture and dilute to 50-100 ml.
16. Plate out at 100 µl per well into sterile 96 well ELISA plates. Plates are stored in CO_2 incubator (10% CO_2 at 37°C).
17. After 3-4 days add 100 µl HT medium (HAT without aminopterin or HT medium).
18. After 7 days remove 180 µl of media and add fresh HT media (equivalent to volume removed)
19. After 10 days (from start of procedure) test hybridoma clones for specific antibody production by ELISA.

THE LOGIC

- Unfused myeloma cells can not grow because they lack HGPRT.
- Unfused normal spleen cells can not grow indefinitely because of their limited life span.
- However, hybridoma cells (produced by successful fusion) are able to grow indefinitely because the spleen cell partner supplies HGPRT and the myeloma partner is immortal.

SCREENING OF HYBRIDOMA FOR ANTIBODY PRODUCTION

For plant viruses usually screening of the clones is done by ELISA using specific antigen and the extract of hybridomas. The steps used in ELISA will be described in the next chapter. Important points are given below:

1. Test the supernatant for each culture to find those producing the desired antibody.
2. Because the original cultures may have been started with more than one hybridoma cell, you must now isolate single cells from each antibody positive culture and subculture them.
3. Test each supernatant for the desired antibodies. Each positive subculture-having been started from a single cell-represents a clone and its antibodies are monoclonal. That is, each culture secretes a single kind of antibody molecule directed against a single determinant on a preselected antigen.
4. Scale up the size of the cultures of the successful clones.
5. Hybridoma cultures can be maintained indefinitely.
6. The yield runs from 10-60 μg/ml when growing in mice. Here the antibody concentration in the serum and other body fluids may reach 1-10 mg/ml.

Expansion and recloning

Two methods are used:

Method-1

1. Of a 96 well culture plate add 100 μl of 20% FCS + RPMI + thymocytes to each well.
2. To each well in the first column add 300-500 cells (10-50 μl from the original).
3. Dilute serially by removing 100 μl from the first column and mix thoroughly by pipetting up and down. Withdraw 100 μl and add to the next column. Repeat the procedure until all columns are serially diluted.
4. Incubate 8-10 days in CO_2 incubator feeding with additional media only after verifying that a single colony exists in some of the wells.
5. Retest clones for antibody production.

Method-2

Using a haemocytometer, accurately count the number of cells present. Based on the calculated number, dilute the original stock so that an average of 50, 20 and 5 cells per ml is attained. Add thymocytes and plate out 100 μl per well. The amount of solution plated out will vary according to your estimate, though a minimum of 48 wells per clone per dilution is recommended. Retest clones for antibody production.

When any given well has 30% of its surface covered with hybrid cells, then dilute contents of the well by a factor of 10. It means if a vigorously growing culture is diluted 10 times it will be back to its original numbers in approximately 2-3 days.

CRYOPRESERVATION

1. Harvest the cells from two 10 ml Petri plates by centrifugation.
2. Resuspend the pellet in 3-5 ml of 20% FCS + RPMI + 10% DMSO
3. Aliquot exactly 1 ml into freezer vials and place them into insulated box and deposit to -80°C freezer for 6-24 h
4. Transfer to liquid nitrogen system.

FACTORS AFFECTING MAb PRODUCTION

1. Purity of the antigen.
2. The mouse stain which is used.
3. The age and sex of the mouse
4. Use of adjuvants.

USES OF MONOCLONAL ANTIBODIES

1. MAbs are used in measuring protein and drug level in serum.
2. Typing tissue and blood.
3. Identifying infectious agents like viruses, their strains and phytoplasmas.
4. Identifying clusters of differentiation for the classification and follow up therapy of leukemias.
5. Identifying tumor antigens and auto-antibodies.
6. Identifying specific cells involved in the immune response. Identifying and quantifying hormones.

PHAGE DISPLAY TECHNOLOGY

Phage display is a method for the study of protein-protein, and protein – DNA interacrions that utilizes bacteriophage (bacterial virus) to connect proteins with the genetic information that encodes them. The most common baceriophages used in phage display are M13 and fd filamentous phage. Its potential applications include selection of high-affinity antibodies, characterization of antibody-binding sequences and investigation of potential ligands for orphan molecules.

In case of M13 filamentous phage display, the DNA encoding the protein or peptide of interest is ligated into the pIII or pVIII gene. The phage gene and insert DNA hybrid is then transformed into *E. coli* bacterial cell such as TGI or XLI-Blue *E. coli.* The phage particles will not be released from *E. coli* cells until they are infected with helper phage, which enables packaging of phage DNA and assembly of the mature virions with the relevant protein fragment as part of their outer coat on either the minor (p. III) or major (p. VIII) coat protein. The incorporation of many different DNA fragments into p. III or p. VIII genes generates a library from which members of interest can be isolated.

By immobilizing a relevant DNA or protein target(s) to the surface of a well, a phage that displays a protein that binds to one of those targets on its surface will remain while others are removed by washing. Those that remains can be eluted, used to produce more phage (by bacterial infection with helper phage) and so produce a phage mixture that is enriched with relevant (*i.e.* binding) phage.The repeated cycling of these steps is referred to as 'panning' in reference to the enrichment of a sample of gold by removing undesirable materials.

Phage eluted in the final step can be used to infect a suitable bacterial host, from which the phagemids can be collected and the relevant DNA sequence excised and sequenced to identify the relevant, interacting proteins or protein fragments. The following protocol is generally used.

1. Target proteins or DNA sequences are immobilized to the wells of a microtitre plate.
2. Many genetic sequences are expressed in a bacteriophage library in the form of fusions with the bacteriophage coat protein, so that they are displayed on the surface of the virus particles. The proteins displayed corresponds to the genetic sequence within the phage.
3. This phage-display library is added to the dish and after allowing the phage time to bind, the dish is washed.
4. Phage-displaying proteins that interact with the target molecules remain attached to the dish, while all others are washed away.
5. Attached phage may be eluted and used to create more phage by infection of suitable bacterial host. The new phage constitutes an enriched mixture, containing considerably less irrelevant (*i.e.* non-binding phage) than were present in the initial mixture.
6. The DNA within the interacting phage contains the sequences of interacting proteins, and following further bacterial-based amplification, can be sequenced to identify the relevant, interacting proteins or protein fragments.

Phage display is a widely used method for *in vitro* protein evolution, also called protein engineering. Competing methods for *in vitro* protein evolution are yeast display, bacterial display, ribosome display and mRNA display.

PHAGE AS ALTERNATIVES TO ANTIBODIES

Phage display technologies can be used to generate high-affinity antibodies to proteins of interst, essentially bypassing conventional immunization and hybridoma technology. Antibody isolation is accomplished by panning a library that displays variable region of immunoglobulin genes. There are several ways to construct a phage-based antibody library, the most common of which is to clone the variable region of both immunoglobulin (Ig) light (vl) and heavy (VH) chain genes upstream of pIII phage coat protein gene. The VL and VH genes are amplified from lymphocytes using the polymerase chain reaction (PCR) and connected via a linker region so that, upon expression, the fusion proteins, also known as single-chain antibodies, resemble the antigen binding region of a traditional antibody. Panning of the library against an antigen of interest allows for isolation of phage displaying single chain antibody that binds to the antigen. Each antibody- displaying phage particle can be amplified and retained indefinitely, similar to a hybridoma cell line for a monoclonal antibody. Additionally, a process analogous to affinity maturation, termed chain shuffling, can further improve the binding affinity of an isolated phage. This chain shuffling process involves the generation of two libraries, one in which the VL gene of the best binding phage is held constant and displayed with a library of VH genes. Panning of both libraries and assembling optimal VH and VL into one molecule may produce antibodies with significantly higher binding affinity. Antibodies with higher affinity are usually more useful in research. Phage-display antibody libraries are particularly useful for the isolation of antibodies that may be difficult by traditional means.

As an alternative to phage-display technique, plasmids and polysomes have also been used. Phage display and other biologically amplifiable display systems have revolutionized the study of protein-protein interactions

SEROLOGICAL DIAGNOSIS OF PLANT VIRUSES/PHYTOPLASMAS

The antibodies are γ-globulins. The main one is IgG with a molecular weight of about 150000. It is Y shaped. The optimum proportion of antibody and antigen gives a visible precipitate. If any one of the components is in excess the process of precipitation may not occur.

Reaction of antibody-antigen only when they are in optimum condition

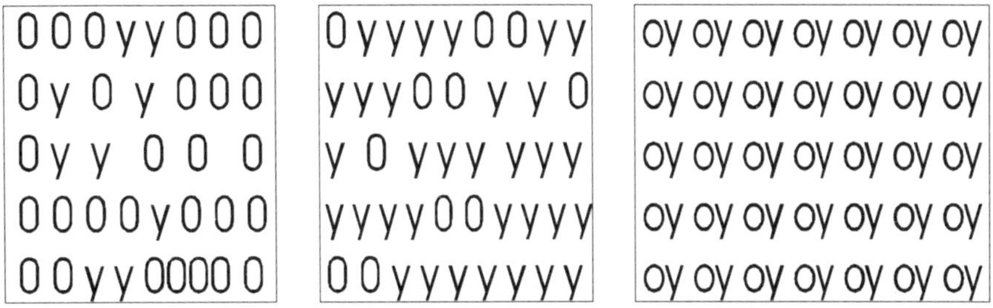

A **B** **C**

A = Excess of antigen (No reaction)

B = Excess of antibody (No reaction)

C = Optimum antibody/antigen (will give a precipitate)

TITRE OF ANTISERUM

The highest dilution of the antiserum that will react with its homologous antigen (virus) is determined in precipitin tube test and also by immunosorbant electron microscopy. The following serological tests have been developed from time to time.

PRECIPITIN TUBE TEST

1. The test is carried in thin walled glass tubes (75 x 9 mm).
2. A constant dilution of antiserum is put in 10 glass tubes and dilutions are made by 0.9% sodium chloride as 1:4, 1:64, 1:128 and so on.
3. Add 0.5 ml of diluted virus preparation or partially purified preparation to 0.5 ml of each antiserum dilution. Shake for sometime and then put at 38°C in a waterbath.
4. Observe each tube after 5-10 min against a black background and record the intensity of the precipitate in each tube.
5. For control, healthy antigen and normal antiserum should be used.

The resulting flocculent precipitate is obtained for filamentous viruses and densely granular for viruses with spherical particles.

PRECIPITIN RING TEST

1. Carefully layer the virus preparation on top of the antiserum in a glass tube (3-6 mm diameter). The antiserum should be made

viscous with the help of glycerol or sucrose (10 and 3% respectively).

2. A ring of precipitate is formed at the interface of two layers if the reaction is positive.

MICO-PRECIPITIN TEST

1. The bottom of a Petri plate is coated with silicon or formvar to avoid spreading of reactants.
2. 10 µl drops of antigen are put at the bottom of Petri plate.
3. Over each drop 10 µl of different dilutions of antiserum are put.
4. The dish is covered to avoid evaporation and incubated.
5. The precipitation is observed at low magnification in a light microscope.

GEL DIFFUSION TEST (MICROPRECIPITIN)

Precipitation may also occur on solid medium such as agar as has been explained in this test.

Radial diffusion:

1. Prepare agar gel at a concentration of 0.7 to 1.0% by usual method and spread evenly in a Petri dish.
2. The gel is allowed to set and a well is cut with the help of a cork borer in the centre and 8 wells around it.
3. Add solution of antiserum to the central well and antigens to outer wells and incubate.
4. If a reaction is positive a halo of precipitation is formed due to diffusion of antibody/antigen from the well. The antigen and antibodies will mix at places around the well and form precipitate which can be seen with naked eyes (Fig. 52).

AGAR DOUBLE DIFFUSION TEST (OUCHTERLONY)

1. Dissolve 7.5 g agar, 9.0 g sodium chloride and 0.4 g sodium azide in 11 ml distilled water and boil.
2. Spread the agar in a Petri dish to a depth of 5mm or if slides are used put 5 ml of melted agar to the slide.
3. Cut wells with the help of a cork borer in such a way that the well first cut in the centre and then around it. If it is on slide cut 4 wells (4 mm diameter each) around the central well (5 mm dia) at

equal distance. If it is in Petri dish then cut 8 wells surrounding the central well 2.5 mm from the edge of central well.

4. Add antiserum to the central well either undiluted or dilute with 0.9% NaCl. The appropriate dilution of antiserum is required to get optimum precipitation line.
5. Purified virus preparation or sap from leaves of infected plant is used as antigen and is put in the peripheral wells. If purified virus is used, it should be diluted upto 1:10 by 0.9% NaCl. If sap is used, it should be homogenized in 1.8% NaCl and the sap is filtered.
6. Control wells with healthy plant sap should be used.
7. Place slide or Petri plates in a sealed plastic container to avoid drying of agar.
8. Leave for 24 h.
9. Examine precipitation lines using an indirect source of light beneath the agar. If photographs are taken, stain with 0.1% 3-4 di-hydroxyphenyl-DL alanine (DOPA). It can be prepared by dissolving 1.97 g DOPA in 100 ml of 0.1M potassium phosphate buffer at pH 7.4.

Note: Long filamentous particles will not diffuse through the agar unless they are broken into small fragments through sonication.

The reaction of precipitation lines in Ouchterlony test will depend upon antiserum and antibody concentrations and may be represented as:

i) If the antigen is filamentous (such as TMV), the precipitation line will be very close to antigen wells.
ii) Small isometric particles will diffuse rapidly and will form precipitate towards the antiserum well.
iii) If the antigen and antibodies are contaminated with healthy host antigen like fraction-1 protein and its antibodies in antiserum, the precipitation will form close to antiserum well.
iv) The unbalanced reactants also result in alter the precipitation line. The zone of precipitation will broaden and move away from the well having excess reactant.
v) Optimum concentration of antibody-antigen will develop a thin distinct precipitation line.
vi) If two serological identical virus isolates are put in adjacent wells, a continuous confluent precipitation line between them and antiserum is formed. Thus two unknown viruses can be identified if they are serologically related.
vii) If, however, a spur (partial fusion) precipitation line is formed between the two virus isolates. A spur reaction is formed when

two viruses share at least one common antigenic protein but not the other antigenic proteins. In this way it can be concluded that two viruses are related but not serologically identical.

AGGLUTINATION TESTS

CHLOROPLAST AGGLUTINATION OR SLIDE AGGLUTINATION TEST

1. A drop of crude sap from infected/healthy plants is put on a microscopic slide and mixed with homologous antiserum.
2. If reaction is positive, the chloroplast and other cell debris clump together. This test is used in field testing of viruses in potato.

LATEX FLOCCULATION TEST

1. The antigen or antibody is absorbed on to polysterene latex particles.
2. Mixed with corresponding antigen or antibody.
3. Visual clumping of latex particles will be observed.

HAEMOAGGLUTINATION TEST

Bentonite, barium sulphate or sheep red blood cells treated with tannic acid have been used for latex agglutination test. These tests are highly sensitive than microprecipitin or immunodiffusion tests and can be carried out with lower concentration of reactants.

ENZYME-LINKED IMMUNOSORBENT ASSAY (ELISA)

In serodiagnostic detection of plant viruses, the IgG fraction of the antiserum is most effectively used in the techniques like ELISA, DIBA, ISEM etc.

ELISA is one of the most widely used serological tests for plant virus detection, identification, quantification and comparison owing to its simplicity, sensitivity (detection limit 1-10 ng/ml) and adaptability. The amount of virus present is proportional to the amount of enzyme-labeled specific antibody forms the basis of the test.

Purification of IgG

1. To 10 ml antiserum, add 9 ml of distilled water
2. To (1), add 10 ml of saturated ammonium sulphate solution drop by drop.

3. Incubate 30 to 60 min at room temperature.
4. Centrifuge at 10,000 x g for 10 min.
5. Dissolve pellet in 2 ml of ½ strength phosphate buffer saline (PBS).
6. Dialyze 3 times against at least 500 ml of ½ strength PBS (4 h each) at 4°C. Use low cut-off dialysis tube.

Preparation of column

Use a Bio-rad 7 mm I.D. Econo column with a stop cock valve attached (A broken 10 ml pipette will also work as column). Add approximately 5-10 ml of DEAE cellulose (DEAE-sephacel, pharmacia chemicals). This comes beaded and pre-swollen and needs limited equilibration.

The DEAE cellulose slurry is poured down by the side of the column to avoid formation of air bubble. Allow the cellulose to settle. The column height should be approximately 7.0 cm.

The column is equilibrated with half strength PBS. When the pH of the ingoing buffer (½ strength PBS) matches the pH of the effluent, then add liquid carefully to side of column so that cellulose is not disturbed once packed. After equilibration, adjust buffer level to top of cellulose in column.

FILTRATION

The salt fractionated (SF) serum is added drop wise to the pre-equilibrated DEAE cellulose column. Run column until SF serum level is at the top of the cellulose in column (start collecting affluent as soon as column is started). Add 10-15 ml of half strength PBS to column carefully and continue collecting effluent until 12 one ml fractions are collected.

EVALUATION

Monitor effluent (each tube separately) at 280 nm on a spectrophotometer with ½ strength PBS as reference.

Combine tubes with OD280 value above 0.8 and adjust strength (with ½ strength PBS) of γ-globulin to read approximately1.4 OD units. If OD of combined fractions is less than 1.4, calculate γ-globulin concentration (1.4 OD = 1 mg/ml).

Store purified γ-globulins in plastic or glass vials at 4°C. Discard the used cellulose and acid wash the column. An elution curve for immunoglobulin from DEAE column is shown in Fig. 53.

CONJUGATION

1. Centrifuge 1 ml (= 5 mg) of enzyme precipitate (Alkaline phosphatase type VII sigma) at 10000 x g for 10 min. Discard supernatant.
2. Dissolve pellet in 2 ml (2 mg) of purified γ-globulins.
3. Dialyze 3 times against 500 ml PBS (4 h each) at 4°C to remove excess ammonium sulfate.
4. Add fresh, clear glutaraldehyde solution to about 0.05% final concentration and mix well.
5. Incubate 4 h at room temperature. A clear, light yellow-brown color should develop.
6. Dialyze three times in 500 ml PBS to remove glutaraldehyde (4 h each at 4°C).
7. Add bovine serum albumin to a concentration of 5 mg/ml and store at 4°C.

ENZYME LINKED IMMUNOSORBENT ASSAY (ELISA)

Several modifications have ben suggested in ELISA technique but most commonly double antibody sandwich (DAS-ELISA) and direct antigen coating (DAC-ELISA) are being used in detection of plant viruses.

DOUBLE ANTIBODY SANDWICH ELISA (DAS-ELISA)

Before starting ELISA one should determine optimum concentrations of coating and enzyme-labeled γ-globulins as per Fig. 54. The steps of ELISA are given below and also shown diagrammatically in Fig. 55.

1. Add 200 μl purified γ-globulins in coating buffer to each well of ELISA plate (1-2 μg/ml).
2. Incubate the plate 1-4 h at 37°C.
3. Wash ELISA plate with PBS-T (three times).
4. Add 200 μl test sample in each well (for citrus, grind 1 part tissue in 10-20 parts PBS-tween + 2% PVP-40 or other suitable buffer.
5. Incubate the plate overnight at 6°C or 2-4 h at 30°C.
6. Wash ELISA plate with PBS-T (three times).
7. Add 200 μl enzyme-labeled (Alkaline phosphatase) At an appropriate concentration to each well (usually 1:500 to 1:5000).
8. Incubate the plate 1-4 h at 30°C.
9. Wash the plate three times with PBS-T.

10. Add 200 µl p-Nitrophenyl phosphate substrate 0.6-1.0 mg/ml in diethanolamine buffer
11. Incubate the plate 0.5-2 h at room temperature
12. Yellow color will develop in positive sample wells (Fig. 56), also take calorimetric measurement at 405 in an ELISA reader
13. Stop reaction with 50 µl 3.0M NaOH

Note: Use reliable negative control for each test in a plate. Absorb antiserum against healthy plant extract. Use fresh substrate and check for spontaneous colour change.

Note: Enzyme alkaline phosphatase is commonly used to make conjugate but other enzymes like penicillinase and horseradish peroxidase have also been used as conjugate. The substrate for penicillinase is penicillin (Fig. 57 shows results of penicillinase-based ELSA).

STOCK SOLUTIONS

A. Coating buffer (pH 9.6)

1.59 g Na_2CO_3
2.93 g $NaHCO_3$
0.20 g NaN_3

B. Stock buffer PBS (pH 7.4)

8.00 g NaCl
0.20 g KH_2PO_4
2.90 g $Na_2HPO_4.12H_2O$ or 45 g anhydrous
0.20 g KCl
0.20 g NaN_3

C. Wash buffer PBS-T

PBS + 0.05% Tween-20

D. Tissue grinding buffer PBST + PVP

1. PBS-T + 2% PVP 40
2. Other extraction buffers, Tris and phosphate buffers 0.05 M to 0.015 M at pH 7.5 to 8.0.

E. Enzyme conjugate buffer (PBS-T + PVP +OA)

PBS-T + 2% PVP + 0.2% ovalbumin

F. Substrate buffer (pH adjusted to 9.8 with HCl)

97 ml diethanolamine

800 ml distilled water

0.20 NaN_3

G. Stopping solution

120 g 3 M NaOH, Distilled water 1 litre

DIRECT ANTIGEN COATING-ENZYME-LINKED IMMUNO-SORBENT ASSAY (DAC-ELISA)

1. Add 200 µl test and control samples extracted or diluted in coating buffer containing 25 (W/V) PVP to each well of the ELISA plate. Cover the plate and incubate at 37°C for ½-1 h.
2. Empty and wash the plate by flooding the wells with PBS-T for 3 min, three times. After final washing drain out all liquid on a paper towel.
3. Add 200 µl blocking solution (make by adding 5.0 g bovine serum albumin (BSA)/spray dried milk to one litre PBS-T) to each well. Incubate at 37°C for 1h to block remaining reactive surfaces.
4. Wash the plate as in 2 above.
5. Add 200 µl crude antiserum (primary antibody) diluted in PBS-TPO (PBS-T polyvinyl-pyrrolidone ovalbumin) to each well. Incubate at 37°C for 2 h or overnight at 4°C.
6. Wash the plate as in 2 above.
7. Add 200 µl alkaline phosphatase enzyme-labeled antirabbit IgG (secondary antibody, commercially available) diluted in PBS-TPO to each well. Incubate at 37°C for 2 h or overnight at 4°C.
8. Wash the plate as above.
9. Add 200 µl freshly prepared substrate (paranitro-phenyl phosphate 0.5-1.0 mg/ml) to each well. Incubate at room temperature for 1-2 h or as long as necessary to observe reaction.
10. Terminate reaction, if required, by adding 50 µl 3 M NaOH to each well.
11. Observe the color intensity by visual observation or by measuring absorbance at 405 nm in an ELISA reader. Positive samples will show absorbance value 3 times more than healthy control. Positive samples show yellow colour after adding the substrate
12. Include both positive and negative reference samples to validate the assay.

DOT IMMUNOBINDING ASSAY (DIBA)

Dot immunobinding assays are also known as dot- blot immunoassay or sometimes referred as DOT-ELISA. This technique is useful because the dotted membrane can be stored and processed at convenience or may be sent to other laboratories for processing if required facilities are not available with the worker.In this technique, antigens are immobilized directly to the nitrocellulose membrane (NCM). The remaining membrane surfaces are blocked by incubation in a solution of a non-reactive heterologous protein, to prevent non-specific adsorption of antibody to the membrane. The procedure is given below:

1. Extract sap from infected tissue in antigen extraction buffer 1:10 (W/V). Filter the sap through muslin cloth.
2. Take 0.8 ml of filtrate in to a 1.5 ml eppendorf tube and add 0.4 ml chloroform, vortex and centrifuge at 12000 x g for 2 min.
3. Take 200 µl of the clarified sap into 800 µl of antigen extraction buffer and vortex.
4. Cut a desired size of nitrocellulose membrane (NCM) and make squires of 1x1 cm with a pencil. (Handle NCM with gloves/forcep)
5. Wet NCM by floating it on TBS and then air dry. Spot the test/ control samples (5-10 µl) to each squire by pipette.
6. Allow NCM to air dry and immerse in blocking solution with gentle shaking for 1h at room temperature to block left over sites.
7. Rinse once in TBS for 10 min.
8. Incubate the NCM with primary antibody (developed against the virus) diluted in TBS-SDM.
9. Wash the NCM three times (10 min each) with TBS.
10. Transfer the NCM into second antibody (enzyme-labeled antirabbit IgG conjugated with alkaline phosphatase – commercially available).
11. Wash the NCM as earlier.
12. Incubate the NCM in substrate solution at room temperature and watch for colour development (takes 5-10 min)
13. Rince the NCM twice (10 min each) in fixing solution and then air dry.
14. Photograph NCM and store in dark.
15. Positive samples will show deep purple colour (Fig. 58).

REAGENTS TO BE USED IN DIBA

Tris buffered saline (TBS, pH 7.5)

0.02 M tris and 0.5 M NaCl are dissolved in water, adjust pH 7.5 with HCl and make up the volume two litres.

Antigen extracting buffer (TBS + 50 mM DIECA)

Add 11.25 g DIECA to 1 litre of TBS

Blocking solution (TBS-SDM)

Add 5.0 g spray dried milk (SDM) to 100 ml TBS.

Antibody-conjugate buffer

same as TBS-SDM.

Substrate buffer (pH 9.5)

0.1 M Tris, 0.1 M NaCl, 5mM $MgCl_2$, dissolve in water and adjust pH 9.5 with HCl and make up the volume upto one litre.

Substrate solution

Solution A

Dissolve 75 mg of nitroblue tetrazolium (NBT) in 1 ml of Dimethyl formamide.

Solution B

Dissolve 50 mg of 5-Bromo-3-Cloroindolyl phosphate (BCIP) in one ml of dimethyl formamide.

Add NBT and BCIP to give final concentration of 0.33 mg/ml and 0.175 mg/ml respectively to the substrate buffer before use.

Fixing solution

10 mM tris, 1 mM EDTA, dissolve in water and adjust the pH to 7.5 with HCl and make up the volume upto one litre.

DIRECT TISSUE BLOT IMMUNOBINDING ASSAY

It is a useful technique for detection of phloem borne viruses infecting woody and perennial crops. Unlike DIBA, the ooze from the infected tissue is directly blotted or impressed on the nitrocellulose membrane by cutting the edge of the twig. The technique is specially useful for mass screening of field trees.

Protocol

1. Cut a desired size of nitrocellulose membrane (NCM) and make squires of 1 x 1 cm with a pencil.
2. Cut the twig repeatedly with a sharp razor blade. After each cut imprint on each NCM squire. If the tissue is more juicy then blot

to the next squire without making another cut. Always use healthy control in test and a few squire may be blotted by the cut of healthy twig.

3. Dry the samples at least for 30 min.
4. Place NCM in a flat container having 1% BSA in PBS (pH 7.5). Place either at 4-6°C for at least 2 h or overnight at room temperature.
5. Wash NCM with PBS-T in a container three times. Agitation is preferred to wash the entire blots equally.
6. Add the monoclonal antibodies (specific to the virus) with 1% normal serum. Place the container on a shaker for 1-2 h at room temperature.
7. Wash three times in PBS-T as in 5 above.
8. Block with 1% BSA for 30 min at room temperature (optional)
9. Add Goat-anti-mouse (Biotinylated) conjugate at the suggested dilution (commercially available). Incubate 1-2 h on the shaker at room temperature.
10. Wash three times with PBS-T as in 5 above.
11. Block with 1% BSA for 20 min at room temperature.
12. Add stept-avidin at the suggested dilution in PBS. Incubate at room temperature for 30 min.
13. Mix the substrate (BCIP and NBY) in 0.1 M Tris pH 9.5 which contains 0.1 m NaCl and 0.005 M $MgCl_2$ (Nitro blue tetrazolium (NBT) is dissolved in N,N'-Dimethylformamide (DMF) at a concentration of 75 mg/ml and Bromo chloro-indolyl phosphate (BCIP) at 50 mg/ml in DMF. Stock the solution in amber bottles and keep refrigerated. Add NBT 0.33 mg/ml and BCIP 0.175 mg/ml to the substrate buffer (0.1 M Tris-HCl, 0.1 M NaCl, 5 mM $MgCl_2$ just before use.
14. Deep staining will develop in phloem tissue if the sample is positive (Fig. 59).

Note: These chemicals are toxic, hence handle carefully.

WESTERN BLOTTING

This technique combines the protein fractionating power of gel electrophoresis with the sensitivity and specificity of solid phase-immunoassay. The basic principle is the transfer of electrophoresed proteins from the gel matrix on to a membrane and rendering them accessable for further analysis by specific immunoprobes. Western blotting is used commonly for detection, identification, quantification and comparison of plant viruses. Following steps are involved in western blotting.

SDS-PAGE

1. Assemble the glass pates as per the manufacturer's instruction.
2. Prepare appropriate volume of resolving gel containing 12% of acrylamide gel.
3. Pour acrylamide solution into the gap between the glass plates and leave enough space for stacking gel. Place the gel in a vertical position at room temperature for polymerization.
4. After polymerization is complete, pour off the overlay and wash the top of the gel with water to remove any un-polymerized acrylamide.
5. Prepare appropriate volume of 5% stacking gel and pour directly on to the surface of the polymerized resolving gel.
6. Insert the comb into the stacking gel solution (avoid air bubbles) Add more stacking gel solution, if required, to fill the spaces of the comb completely. Allow the gel to polymerize at room temperature.
7. Remove the comb after the polymerization and wash wells with water to remove un-polymerized acrylamide, if any.
8. Mount the gel in the electrophoresis apparatus and add tris-glycine electrophoresis running buffer to the top and bottom tanks.
9. Prepare the samples by heating them to 100°C for 3 min in sample buffer to determine the proteins. Load (upto 20 µl) each sample in a pre-decided order into the bottom of the wells.
10. Attach the electrophoresis apparatus to an electric power supply and run the gel at a constant voltage (100 V) till the bromophenol blue (tracking dye) reaches the bottom of the resolving gel.

Electro-blotting

1. Near the completion of the SDS-PAGE run, wear gloves and cut 4 strips of Whatman-3 filter paper and one strip of nitrocellulose membrane (NCM) of the exact size of the SDS-polyacrylamide gel and mark left corner of the filter.
2. Remove gel from the apparatus and soak in transfer buffer. Also float filter papers and membrane strips on the surface of transfer buffer for 5 min.
3. Assemble "transfer sandwich" under buffer by placing SDS-PAGE gel on the NCM and sandwiching these between filter paper strips, sponge pad and the sandwich holder.
4. Transfer the "transfer sandwich" to the blotting apparatus containing transfer buffer with NCM on the anode side of the gel.
5. Run at a constant voltage (20 Vo for 1-2 h).

Immunological probing

1. After electro-blotting for 2 h, de-assemble the transfer-sandwich layer one by one and transfer the blot to a small tray containing block solution (gel can be stained with coomassie blue to check whether transfer is complete). Leave the blot in blocking solution with gentle shaking for 1 h at room temperature or at 4°C overnight.
2. Wash the blot three times in TBS for 10 min each at room temperature.
3. Incubate the blot in the primary antibody (against target protein) diluted in TBS for 1-4 h at room temperature with gentle shaking.
4. Wash the blot in TBS as in 2 above.
5. Transfer and incubate the blot in the second antibody conjugated with alkaline phosphatase diluted in TBS with gentle shaking for 1-4 h at room temperature.
6. Wash the blot as in 2 above.
7. Submerge the blot in substrate prepared in substrate buffer (0.33 mg nitro blue tetrazolium (NBT) and 0.175 m of Bromo-chloro Indolyl phosphate per ml of substrate buffer).
8. After 20 min stop the reaction by transferring the blot to stop solution before background color becomes significant. Rinse twice in stop solution.
9. Air dry the membrane and store under folds of blotting paper at room temperature (protected from light). The blot may be photographed either wet or dry.
10. The time for adequate color development will vary with concentration of reactants, amount of antigen present. Normally good colour develops within 5-10 min.
11. Binding between antigen in the membrane and enzyme-labeled antibody results in the production of distinct band at the site of antigen (Fig. 60).

Reagents

SDS-PAGE Resolving and Stacking gel composition

Solution	**Resolving gel**				**Stacking gel**
Components	8%	10%	12%	15%	5%
Water	4.6	4.0	3.3	2.3	3.4
30% acrylamide mix	2.7	3.3	4.0	5.0	0.83
1.5 M Tris (pH 8.8)	2.5	2.5	2.5	2.5	0.63*
10% SDS	0.1	0.1	0.1	0.1	0.05
10% ammonium persulphate	0.1	0.1	0.1	0.1	0.05
TEMED	0.006	0.004	0.004	0.004	0.005
Total	**10 ml**	**10 ml**	**10 ml**	**10 ml**	**5 ml**

·Use 1.0M tris (pH 6.8)

30% acrylamide mix is prepared by dissolving 30 g acrylamide and 0.8 g N, N' methylene bisacrylamide in distilled water, mixture filtered through Whatman paper No.1, volume made upto 100 ml and stored in amber colored bottles.

Note: Always bear gloves and mask to handle these chemicals.

Sodium dodecyl sulphate (SDS): A 10% stock solution is prepared in distilled water and stored at room temperature.

Ammonium persulfate: Freshly prepared 10% stock solution in water is used.

TEMED: Use electrophoresis grade TEMED.

Resolving gel buffer (1.5M tris-HCl, PH. 8.8): Dissolve 18.15 g of tris in 80 ml water, adjust pH to 8.8 with HCl and make up volume to 100 ml.

Stacking gel buffer (0.5 M tris-HCl, pH 6.8): Dissolve 6.09 of tris in water, adjust pH with HCl and make up volume to 100 ml.

5X Tris-Glycine electrophoresis running buffer (pH 8.3): Dissolve 25 mM tris base, 250 mM Glycine, 0.1% SDS in water, adjust pH to 8.3 with HCl and make the volume to 1000 ml.

Sample buffer: Distilled water 4.0 ml, 0.5 M tris-HCl (pH 6.8) 1.0 ml, Glycerol 0.8 ml, SDS (10% W/V) 1.6 ml, 2-mercaptoethanol 0.4 ml, Bromophenol blue 0.2 ml. Dissolve the sample in sample buffer at 1:1 (v/v) and heat in boiling water for 3 min.

Blotting/transfer buffer (pH 8.3): Dissolve 39 mM tris base, 48 mM Glycine in 20% Methanol. Make up the volume with distilled water to 3000 ml.

Tris buffered saline (TBS, pH 7.5): Dissolve 0.02 M tris base and 0.15 NaCl in water, adjust pH to 7.5 and make up the volume with distilled water to 3000 ml.

Blocking solution: TBS containing 5% BSA/ skimmed milk.

Substrate buffer (pH 9.5): Dissolve 0.1 M tris base, 0.1 M NaCl, 5 mM $MgCl_2$, in water, make up the volume to 1000 ml and adjust the pH to 9.5.

Substrate: As in DIBA

Stop solution (pH 7.5): Dissolve 10 mM tris base, and 1 mM EDTA in water and make up the volume to 1000 ml and adjust pH to 7.5.

IMMUNOFLUORESCENCE (IF)

The epitopes of the antigen are recognized by the antigen-specific antibodies (poly or monoclonal). In IF, there is a specific binding of antigen and antibody which can be determined in sections or plant sap of virus or phytoplasma infected plants using Fluorescein isothiocynate (FITC) labeled goat anti-mouse/ immunoglobulins produced in rabbit. The test is used in two ways.

1. To detect pathogens (phloem-limited viruses, phytoplasmas and fastidious prokaryotes).
2. To locate the pathogen in plant tissue.

PROCEDURE

Detection of pathogen in sections:

1. Make 3-4 mm long pieces of midribs of leaves of test samples (infected and healthy).
2. Cut 10 µm-thin sections in a freeze microtome in ice blocks.
3. Select sections having intact vascular bundles under a binocular microscope and place them in a cavity slide. Use different slides for infected and healthy sections.
4. Incubate slides at 55°C for 90 min.
5. After incubation, cover each section with a drop of antiserum (specific).
6. Incubate at room temperature for 1 h.
7. Wash thrice in PBS-T using a Pasteur pipette.
8. Drop 15 µl of conjugate [1 ml PBS + 10 µl anti-mouse/rabbit IgG conjugated with FITC (Sigma chemicals) + 10 µl Evans blue] over each section.
9. Incubate in dark for 30 min.
10. Mount in glycerine 1:1 (v/v) in PBS
11. Observe under fluorescent microscope at X40 mag.
12. Apple green fluorescence will be observed in phloem tissue of infected sections (Fig. 61). Healthy sections will show dark red color.

Detection of citrus greening bacterium in plant extract:

1. Remove midrib of 5 leaves from infected plant and five from healthy plant as control.
2. Chop these midribs separately with a blade in Petri plates in 2 ml PBS-600.

3. Remove the extract from Petri plates with the help of a syringe and put separately in an eppendorf tube.
4. Centrifuge 14000 rpm for 10 min.
5. Dissolve the pellet in 200 µl of antigen specific monoclonal antibodies.
6. Incubate at room temperature for 40 min.
7. Add 1 ml PBS after incubation and centrifuge as in 4 above.
8. Dissolve the pellet in 20 µl conjugate (same as has been mentioned for sections)
9. Incubate at room temperature for 30 min.
10. Add 1 ml PBS and centrifuge as in 4 above.
11. Emulsify the final pellet in 2 µl of glycerol and mount on a microscopic slide.
12. Observe under fluorescent microscope at X100 mag.
13. Typical round and elongated forms of greening bacterium with apple green fluorescence will be visible in infected samples but not in healthy (Fig. 62).

Reagents

PBS-600 pH 7.4 (for 500 ml): Dissolve 7.828 g NaCl, 0.09526 g KH_2PO_4, 1.4325 g $Na_2HPO_4.12H_2O$, 0.096915 g KCl in 500 ml water.

PBST: PBS + 0.05% Tween 20.

HPLC PEPTIDE PROFILING

The coat protein peptide profiling reflects the extent of amino acid sequence identity between two proteins and has been shown to clearly differentiate their strains specially in potato Y group of viruses.

Procedure

Isolation of coat proteins:

1. Disrupt purified virus particles in 60% formic acid (v/v) at ambient temperature for 1 h.
2. Centrifuge the suspension at 10,000 g for 3 min.
3. Dialyse three times against distilled water and freeze-dry.

Tryptic digestion of coat proteins:

1. Suspend freeze-dried coat proeins on 0.05 M NH_4HCO_3 (500 µl/mg coat protein).

2. Sonicate 4-5 s to dissolve protein and digest at 37°C for 16-17 h with trypsin at an enzyme-protein ratio of 1:50.
3. Vacuum-dry the digested protein in a speed-vac and dissolve again in 0.1% trifluoroacetic acid (TFA, 500 µl/mg of protein).
4. Centrifuge at 10000 g for 3 min to remove insoluble material.

Isolation of peptides:

1. Load aliquots of soluble tryptic peptides (minimum 50 µg) from coat protein digests on to a C18 reverse phase column (218 TP 10 µm, 4.6 mm x 30 cm, Bondapack, Water Associates) connected to a Maxima 820 chromatography system (Water Associates).
2. Separate components in 0.1% TFA at a flow rate of 1 ml/min using a linear gradient of 0-35% acetonitrite over 70 min at ambient temperature.
3. Detect peptide eluting from the column by a UV monitor at 214 nm.

Amino acid analysis:

1. Hydrolyse coat proteins or peptide fragments obtained from HPLC at 110°C in 5.8 M HCl containing 0.01% phenol for 20-22 h under nitrogen.
2. Analyse samples on Water's Amino acid Analyser using an ion exchange column.
3. Peptide profiles of strains of the same virus are superimposable whereas those of distinct potyviruses are different. Further, profiles of strains of same virus will have 57-100% peaks with identical retention time. In contrast, profiles of distinct potyviruses will have only 16-42% peaks with identical retention times.

The equipments, HPLC, speed-Vac, Amino Acid Analyser will be required for conducting these tests. Chemicals like, acetonitrite, ammonium bicarbonate, formic acid, isopropyl alcohol, methanol, trifluoro-acetic acid and trypsin will also be required.

CHAPTER 8

IMMUNOSORBENT ELECTRON MICROSCOPY (ISEM)

In immunosorbent electron microscopy, serology and electron microscopy are coupled together. In this procedure, antibody-antigen is viewed in Electron Microscope. It is a very useful technique for determining the specificity of the antiserum, identification of viruses and to establish their serological relationship. The principle is that the single virus particle (antigen) has large number of binding sites whereas an antibody has only two binding sites. Therefore, when antigen and antibody are brought together, large number of antibodies binds to the antigen which can be seen in an electron microscope (EM) as a black coat around the virus particle and referred as decoration. However, if an EM grid is coated with antibodies and then put to virus preparation, a comparatively large number of particles are trapped. This process is termed as Trapping.

Trapping

a. Float the carbon coated grids (carbon side down) over a drop of 10 µl of diluted antiserum (1:1000) in 0.07 M phosphate buffer (pH-6.5). The drop can be put on a piece of dental wax or parafilm placed in humid petriplate and leave for 1 h at 37°C.

b. Wash the grids for 10-15 min in phosphate buffer.

c. Drain briefly and then place them over a drop of 10 µl virus extract and leave for 30 min at room temperature. For control fresh grids (without putting on antiserum) are put on virus extract.

d. Remove the grids and wash approximately 20 drops of DDW.

e. Stain with 2-4 drops of 2% freshly prepared uranium acetate (UA) solution.

f. Drain the grid to dry.

g. Examine the grids under EM at 20,000 magnification.

h. Count the number of particles from 10 different areas of the treated and control grids. The antiserum coated grids will have much higher couceutration of virus porticles (Fig. 63).

Decoration

Decoration is a process whereby virus particles are first individually adsorbed to grids, and then the grids are floated over specific antibody, a layer of specific antibody halo is seen to have absorbed the stain and became electron dense.

1. Virus particles are adsorbed to the grid.
2. Grid is carefully washed with phosphate buffer (pH-6.5, 0.07 M).
3. Put the grid over a drop of diluted antiserum (1:50) at room temperature for 30 min.
4. Wash the grids for 10-15 drops of DDW and stain with 2% UA and allowed to dry and then examine in EM.
5. All the particle will show decoration if the preparation has only one virus (Fig. 65). However, if two viruses of similar morphology are there in the preparation, then particles of the virus of which antiserum is used will be decorated and others will remain undecorated (Figs. 64a and b).

Gold labelling

It is also a technique when serology is coupled with electron microscopy. Gold labelling is useful in cases where virus particles are with membrane and difficult to decorate *e.g.* Tospovirus or rhabdoviruses. For such viruses decorating antibody with conjugated gold can offer clear diagnosis.

Protocol

- Prepare the grid as described in trapping and rinse carefully with 0.1 M phosphate buffer, pH-7.
- Put the grid for 1 h at room temperature (RT) on 1% BSA in 0.1 M phosphate buffer pH 7.0 (BSA-PB) for blocking.
- Put the grid for 15 min on specific antibody diluted in BSA-PB.

- Rinse with BSA-PB; transfer the grid to an antibody conjugated with gold (commercially available) diluted in BSA-PB for 30 min to 1 h.
- Rinse with PB followed by DDW, and stain with 1% UA and examine in E.M.
- The virus particles will show decoration with gold particles.

In leaf-dip preparation only a few particles will be visible as against large number of particles in trapping. In decoration and gold labelling, the particles will show decoration with antibody and gold particles along the virus when antibody conjugated with gold was used.

Materials required

Clean glass petridish, Parafilm or Dental wax piece, Filter paper, Copper grids 400 mesh size with support film of carbon, Sorensens phosphate buffer, 0.07 M, pH-6.5, Micropipette with disposable tips, Clean and dry twizzers with very fine tips, Negative stain *e.g.* 2% UA (Uranyl acetate) in DDW, Eppendorf tubes 0.5 ml capacity, Minidrill (electric) and wooden tooth picks, Carborundum (600 mesh) powder, Antiserum, ELISA plates, Goat anti-rabbit conjugated with gold 10 nm size, BSA.

CHAPTER 9

NUCLEIC ACID BASED DIAGNOSIS

Nucleic Acid Hybridization

Similar to dot blot immunoassay for detection of virus protein, this test is used for the detection of virus nucleic acid. Detection of very small amount of virus in plant material is possible by nucleic acid hybridization tests. The principle involves the hybridization of labeled probes to cellulose nitrate bound nucleic acid that were spotted directly or transferred after fraction by electrophoresis on agarose gel (Southern transfer). Depending on the extent of homology between the viral nucleic acids immobilized on the cellulose nitrate and the nucleic acid in the probe, hybridization is retained which can be assessed by autoradiography. In other words, the nucleic acid from plant material is irreversibly bound to a nitrocellulose membrane (NCM) or nylon membrane (NM), and is allowed to hybridize with radio-labeled probe to a virus genome suspected to be present in the sample. Positive hybridization indicates the presence of suspected virus in the test samples. This test is also used for detection of phytoplasmas, viroids and fastidious bacteria like citrus greening bacteria.

Southern Transfer

This technique of transferring the DNA from the gel to a NCM/NM was devised by Southern (1975). In this technique, the DNA is transferred from the gel after being denatured and neutralized, through the buffer by capillary action onto the membrane, immobilized and used for hybridization. Once transfer has been done to the membrane, it becomes a replica of the gel. The DNA should be denatured to make it single stranded, to allow the probe to bind to it.

PROTOCOL

1. Trim off the agarose gel above the loading wells, the regions not having any DNA to fix the orientation, make a slanting cut on the left hand corner.
2. Take measurement of the gel.
3. Soak the gel for 10 min in several volumes of 0.2 N HCl. This step is called depurination. By doing this, DNA is partially hydrolysed, can pass through the gel even if it gets dried. The resulting fragments are transferred efficiently. Avoid depurination for longer time.
4. Rinse it in deionized water.
5. Soak the gel in a tray with denaturation solution with constant gentle agitation.
6. After 45 min, pour out the denaturation solution and add neutralization solution. Soak the gel for 30 min with constant gentle agitation.
7. While the gel is soaking, cut the membrane as per the gel measurement, trim off the membrane on the upper left hand corner. Handle the membrane with gloves and forceps. Do not touch it by hand.
8. Soak the membrane in sterile distilled water (SDW) for 2 min.
9. Soak the membrane in 10 x SSC for 15-20 min.
10. Pour transfer buffer in a tray and set the transfer. Keep a glass plate across the tray and keep a 3 M Whatman filter paper as a wick. Reverse the gel and keep it on the wick. Put the membrane on the gel carefully. Roll a pipette over it to remove air bubbles. The left cut corner of the membrane should coincide with that of the gel. Put three sheets of Whatman paper cut to size. Put stacks of blotting paper on it. Every time take care to avoid air bubbles. Put 500 g-1 kg weight to give pressure. Allow the transfer to continue overnight.
11. Next day, remove the assembly with forceps.
12. Soak the membrane I 6 x SSC for 5 min to remove agarose pieces.
13. Dry the membrane. Immobilise the DNA by baking at 80°C for 2 h. If nitrocellulose membrane is used, it must be baked in vacuum.

HYBRIDIZATION WITH A PROBE

1. Denature the DNA at 100°C for 5 min. Immediately mix the following:

Denatured DNA	1 µl (30-50 ng of DNA)
Primer (Random hexanucleotides)	1 µl
10 x labeling buffer	5 µl
dNTP (10mM dATP, dGTP, dTTP)	2 µl
Dithiothreitol (DTT) (20 mm)	2.5 µl
Nucleage free BSA	2 µl
(a32P) dCTP (4000 ci/mmol)	1 µl
Klenow (DNA Pol I) (3U)	1 µl
SDW (distilled water)	34.5 µl
	50 µl

Incubate at 37°C for 2 h.

PROBE PREPARATION

The DNA is labeled by Random primer labeling. Probe is made by random primer labeling method as given below. To the DNA fragment random primers are added with one radioactive (dATP or dCTP) and three other non-radioactive dNTPs. Add the following to the eppendorf tube in the order given below:

Water	33 µl
Oligolabeling buffer	10 µl
Bovine serum albumin	2 µl
Template DNAs	1 µl (20 ng of DNA/ml)
32_P-dCTP	3 µl (10 m Ci)
Klenow enzyme	1 µl
Total	**50 µl**

Incubate behind the lucit screen at room temperature for 1 h.

Stop the reaction by adding 2 µl of 1 mM EDTA.

In the presence of Klenow, new strands are synthesized which will have incorporation of radioactive dNTP.

PRE-HYBRIDIZATION

1. The spaces in the membrane where DNA is not bound should be blocked to avoid non-specific binding of the probe. This is done by

immersing the membrane in a solution containing large molecular weight DNA or RNA alongwith detergent and dispersing agent. The yeast mRNA acts as a blocking agent.

2. Keep the membrane in one of the large glass cylinders of a hybridization oven. Balance with another cylinder. Add prehybridization solution in both the cylinders. The SDS in the prehybridization solution reduces non-specific probe binding and prevents nuclease activity. Cylinders are rotated to facilitate uniform spreading of the solution.

HYBRIDIZATION

1. After 4 h, add the probe into the prehybridization solution. Before adding the probe, it should be denatured by boiling (poke a vent hole in the tube before putting it in the water bath). Normally, probe is added at 0.5 x 10^6 cpm/ml of hybridization solution.
2. Incubate over night at 65°C.

USEFUL FACTS

1. A combination of temperature and salt concentration should be chosen properly for hybridization.
2. The washing condition should be stringent when temperature of washing is >65°C and salt concentration is very low (0.2 SSC). The temperature and salt concentration should determine through experiments. Higher the sequence homology between the probe and immobilized DNA. The pairing will be retained even at high temperature and low ionic strength of washing. This fact is used in finding out relatedness between different DNA fragments.
3. Nucleic acid hybridization is also used in the detection and diagnosis of virus and virus like pathogens.

WASHING THE BLOTS

1 The next day morning pour out the hybridization solution and add 30 ml washing solution. This is done to remove any unbound probe

2. Rotate the cylinders for 15 min at 65°C.
3. Repeat washing two more times.
4. Take the membrane out by forceps and dry it on paper towels.

DETECTION BY AUTORADIOGRAPHY

The particle emitted from a radioactive atom activates a photographic emulsion, much like that of light or X-rays. If the photographic emulsion is brought into close contact of a radioactive source, the particles emitted by the source leave, tiny, black silver grains in the emulsion after photographic development.

1. Wrap the membrane in a cling film and put it in a X-ray film cassette. Put the membrane in between the intensifying screen to intensify the signal.
2. In dark orient the X-ray film by cutting off at the upper left hand corner and press it against the membrane.
3. Keep at -20°C for 24-48 h.
4. After 24-48 h develop the X-ray film in dark as per the manufacturer's recommendation to detect hybridization pattern.
5. Dark spots or signal in the X-ray film indicate positive hybridization showing the relationship between the probe used and the virus.

DOT-BLOT HYBRIDIZATION or DNA-DNA HYBRIDIZATION

In this technique, the nucleic acid from infected and healthy plants is irreversibly bound to a nitro-cellulose membrane (NCM) or nylon membrane (NM) and is allowed to hybridize with radio-labeled probe of a virus suspected to be present in the sample. Positive hybridization indicates the presence of suspected virus in the test samples. The test used for viruses having DNA genome, phytoplasmas and fastidious bacteria.

PROTOCOL

Blotting

1. Put on disposable gloves and cut a piece of nitrocellulose membrane NCM or nylon membrane (NM).
2. Draw a lattice of 1 cm squire with a soft lead pencil on the membrane and cut one corner to determine orientation.
3. Wet NCM in water, then in 20 × SSC for 30 min.
4. Drain off fluid and spread the membrane on a sheet of Whatman-3 mm filter paper.
5. Ground 1 g of leaf material of each sample in 1 ml of TE buffer. Filter through muslin cloth.
6. Take 200 µl of each extract and denature DNA by adding 50 µl 0.5 N NaOH. Leave at room temperature for 10 min.
7. Add 25 µl 3 M sodium acetate (pH 5.0) to each sample.
8. Make dilution of 1:10 and 1:100 in TE.
9. Carefully dot 2.5-5 µl of samples in each squire on the membrane.
10. Note the format in which samples have been dotted.

11. Always keep a negative (healthy) sample and a positive (known sample) in the test for comparison.
12. Dry the membrane at room temperature and bake in vacuum at 80°C for 2 h.

PRE-HYBRIDIZATION

1. The baked NCM blots are kept in hybridization cylinder.
2. Prehybridization solution was added at the rate of 0.2 ml/sq cm.
3. The cylinders were then incubated at 65°C for 4 h in hybridization oven with gentle rotation.

To make 10 ml prehybridization solution you will require:

Water	6.3 ml
50 x denhardt's solution	0.8 ml
10 mg/ml sonicated Salmon sperm DNA	0.1 ml
20 x SSC	2.5 ml
10% SDS	0.3 ml

Heat at 100°C in boiling water bath for 90 sec. Put into the cylinder (bag) and seal. Take care to avoid bubbles. Incubate in shaking water bath at 65°C overnight.

HYBRIDIZATION

Add probe at the rate of 1 x 10^6 cpm/ml of solution. Denature probe by boiling for 5 min in water bath and then add to the bag. Incubate in shaking water bath at 65°C overnight.

Washing and Autoradiography

1. Take out the blot by cutting the bag carefully.
2. Discard the solution with the radioactive probe in to radioactive waste.
3. Transfer the blot to a plastic tray filled with washing solution.
4. Wash with 2 x SSC with 0.1% SDS at 65°C on a shaker with moderate agitation. Each washing should continue for 15-20 min.
5. After two washing blot off excess liquid. Dry it on paper towel placed between two layers of thin polythene sheet or sling film in a cassette with intensifying screen.
6. In a dark room, place X-ray film on top, close cassette and place at -70°C overnight.

Observation

Dark spots or bands in the X-ray film indicate the positive hybridization. The intensity of spots will depend on the time of exposure to X-ray film. The negative control will not show any reaction (Fig. 66). These radioactive probes can also be used for detection of pathogen in insect vector (Fig. 67).

Reagents used in nucleic acid hybridization tests

20 x SSC

NaCl	175.3 g
Sodium citrate	88.2 g

Dissolve in 800 ml water. Adjust the pH to 7.0 and make up the volume to one litre. Dispense in aliquots and sterilize.

TE buffer (pH 8.0), 10 mM Tris, 1mM EDTA

1. Dissolve 121.1 g of tris in 800 ml water. Adjust the pH to 8.0 with concentrated HCl. Make up the volume to one litre. Dispense into aliquots and sterilize. It is now ready 1 M tris pH 8.0.
2. Dissolve 186.1 g of EDTA in 800 ml water, adjust pH to 8.0 with 10 N NaOH. Make up the volume to one litre. Dispense into aliquots and sterilize. It is 0.5 M EDTA of pH 8.0.
3. Make appropriate dilutions to prepare required amount of 1 x TE.

10% SDS

Dissolve 100 g of Analar grade SDS in 900 ml water. Heat the mixture at 68°C to dissolve. Adjust pH to 7.2 by adding few drops of conc. HCL. Adjust the volume to one litre with water. Dispense in aliquots.

0.5 N NaOH

Dissolve 20 g of NaOH in 800 ml water and make up the volume to one litre.

3 M Sodium acetate, pH 5.0

Dissolve 408.1 g of sodium acetate $3H_2O$ in 800 ml of water. Adjust the pH to 5.0 with glacial acetic acid. Adjust the volume to one litre and autoclave.

Bovine serum albumin (fraction V sigma)

Dissolve at the rate of 10 mg/ml, filter and sterilize aliquots. Store at -20°C.

5 x Oligolabeling buffer

250 mM tris HCl (pH 8.0)

25 mM $MgCl_2$

5 mM B-mercaptoethanol

2 mM each ATP, GTP, TTP

1 M HEPES (adjusted to pH 6.6 with 4 N NaOH)

1 mg/ml oligonucleotides

Stop buffer (2 mM EDTA)

Take 4 ml of 0.5 M EDTA in water and dilute to 1000 ml.

50 x Denhart's reagents

5 g Ficoll (type 400)

5 g Polyvinyl pyrrolidone (PVP)

5 g Bovine serum albumin

500 ml Water

Salmon sperm DNA

ssDNA is dissolved in water at a concentration of 10 mg/ml. The solution is stirred for 2-4 h at room temperature. The concentration of NaCl is adjusted to 0.1 M and the solution is extracted once with phenol, once with phenol : chloroform. DNA is sheared by passing 12 times through a 17 gauge hypodermic needle. The DNA is precipitated by ice cold ethanol. It is recovered by centrifugation, redissolved at a concentration of 10 mg/ml. Boil the solution for 10 min, aliquote and freeze at -20°C.

Alternatively, DNA can be prepared as given below and approximately 2.5 µg of DNA can be dotted on the membrane.

1 g of leaf tissue is ground in liquid nitrogen, and extracted in 2% CTAB buffer (100 mM tris, pH 8.0, 200 mM NaCl, 10 mM EDTA, pH 8.0, 2% CTAB). The suspension is incubated at 65°C for 1h and deproteinised by emulsifying with chloroform: isoamylalcohol mixture (24:1). DNA in the supernatant is precipitated by adding N/10 NaOH, pH 4.8 and 0.8 volume isopropanol. After 70% alcohol wash, DNA pellet is dried, dissolve in 1 x TE and estimate concentration.

DNA-RNA HYBRIDIZATION

Total RNA isolation

To detect a RNA virus in infected plants by nucleic acid hybridization, total RNA is isolated from the infected material and is probed with the cDNA clone of the virus of interest. However, in most of the RNA viruses group specific conserved region of the clone may be used. The acid-guanidium phenol chloroform method of total RNA isolation is described below:

1. 0.5 g tissue is minced on ice and homogenized with 5 ml of total RNA extraction buffer in a pestle and mortar at room temperature.
2. Transfer the homogenate to a centrifuge tube and add the following reagents in sequence and mix thoroughly after each addition.

   0.5 ml 2 M sodium acetate pH 4.0

   5.0 ml phenol (water saturated)

   1.0 ml chloroform: isoamyl alcohol (49:1)
3. Shake vigorously for 10 sec and cool on ice for 15 min.
4. Centrifuge at 10,000 g at 4°C for 20 min.
5. Transfer aqueous phase at -20°C for 1-2 h to a fresh tube and mix with 5 ml of isopropanol and incubate.
6. Centrifuge at 10,000 g for 20 min and dissolve pellet in 1.5 ml of total RNA extraction buffer and again precipitate with 1 ml of isopropanol at -20°C for 1 h.
7. Resuspend the RNA pellet in 70% alcohol, vacuum dry and dissolve in 50 µl of 0.5% SDS at 65°C for 10 min. This RNA preparation is used for dot-blot and Northern hybridization analysis.

ANALYSIS OF RNA THROUGH DENATURING (FORMALDEHYDE AGAROSE GEL)

Preparation of denaturing agarose gel

Prepare 1% agarose gel in 10 x MOPS buffer. Add 5.4 ml of 37% formaldehyde and pour in casting tray under fumehood. Run using 1 x MOPS buffer.

RNA sample preparation

Required quantity (20-30 µg) of RNA samples are vacuum evaporated till the volume become 2 µl in each case. Then 10 µl of loading buffer is added and sample is boiled for 2 min before loading on the gel. Electrophoresis is done at 100 V for 1-2 h and viewed on a UV transilluminator after staining with ethidium bromide.

Transfer of denatured RNA to nitrocellulose membrane

RNA fractions resolved on gel were transferred to nylon membrane by capillary method as described by Sambrook *et al.* (1989). After electrophoresis, the gel was soaked in two changes of 20 x SSC to remove the formaldehyde. RNA fragments were then transferred from the gel to nylon membrane by capillary method using transfer buffer (20 x SSC) for 24 h.

Prehybridization

The baked filter (NM) was floated on 6 x SSC until it was thoroughly wet and then submerged for two min. The wet filter was slipped into a hybridization tube. Prehybridizing solution was added at the rate of 0.2 ml for each squire centimeter of nitrocellulose filter and incubated at 65°C for 4h with slow rotation.

Hybridization

The 32p labeled double stranded DNA probe prepared from cDNA clone of interest as described earlier was denatured by heating the probe at 100°C for 5 min. The denatured probe was added to the prehybridization solution and incubated for 18 h at 65°C.

After incubation, the hybridization solution was poured off and the filter was removed and washed in 2 x SSC, 0.1% SDS at room temperature or at 65°C four times, each time for 15 min duration. Then the filter was dried in the fold of tissue paper and exposed to X-ray film (INDU QX-16) in cassettes with intensifying screen for 3-12 h at -70°C.

For dot-blot hybridization, 10 µg each of total RNA isolated from the plants are spotted on NCM or NM. Prehybridization and hybridization steps are similar as described above.

Reagents

Total RNA extraction buffer

Guanidium thiocyanate	4 M
Sodium citrate	25 mM
Sarcosyl	0.5%
2-mercaptoethanol	0.1 M

Loading buffer (for RNA samples)

Formamide	720 µl
10 x MOPS buffer	160 µl
Formaldehyde (37%)	260 µl
Sterile water	100 µl
Ethidium bromide (10 mg/ml)	100 µl
Sterile glycerol	80 µl
Saturated bromophenol-blue in sterile water	80 µl

10 x MOPS buffer (pH 6.0-7.0)

MOPS buffer	0.2 M
Sodium acetate	0.05 M
EDTA	0.01 M

NON-RADIOACTIVE PROBES FOR VIRUS DETECTION

In view of the health and environmental hazards by using radioactivity, nucleic acid technology has led to the development of non-radioactive labeling methods for detection of DNA or RNA by hybridization (Fig. 68). The two methods commonly used are given below:

1. The incorporation of biotinylated nucleotides (*e.g.* bio-11-UTP) in to DNA by nick translation which is then used as a probe in a hybridization reaction. The biotinylated DNA is then detected by incubation with avidin or streptavidin which has been chemically coupled to an enzyme like phosphatase, luciferase, peroxidase etc. Streptavidin-enzyme complexes bind the biotinylated DNA which is then visualized by development of a color reaction. As little as 2.5 pg target sequences can be detected by this method.
2. The covalent linkage of an enzyme to single stranded DNA can be used directly as a hybridization probe. The hybridization reaction can be detected by development of a color reaction. The advantages of using non-radioactive probes for virus detection are:

a) There is no hazard associated with non-radioactive probes.

b) These probes are stable and their use is not restricted by a short half life like radioactivity.

c) Higher probe concentrations can be used to achieve hybridization reactions. Thus increasing sensitivity without causing high background problems.

d) The detection is rapid since no autoradiography is involved.

e) The reagents are inexpensive.

f) Most of the laboratories can use these probes as no facilities are required like that of with radioactivity (such as radioactivity supply and disposal etc.).

These probes are now widely used in molecular biology for detection and quantifying specific sequences.

Note: The non-radioactive kits are now available commercially.

POLYMERASE CHAIN REACTION

Detection methodologies are no longer limited to a portion of genome of viruses, for example the coat protein, but the entire genome. Among the detection technologies developed so far, the polymerase chain reaction (PCR) for DNA or reverse transcriptase (RT)-PCR for RNA has emerged as an effective method for the diagnosis of various types of causal agents. PCR is an *in vitro* method for producing large amounts of specific DNA fragments of defined length and sequence from small amount of template. Advances in the elucidation of virus and viroid genome structure and the availability of their genomic sequences have made it possible to create primers for specific RT-PCR.

Polymerase chain reaction is a primer mediated enzymatic amplification of a DNA segment lying between two regions of known sequences. Two oligonucleotides are used as primers, which on binding the template are extended with DNA polymerase in the presence of dNTPs and Mg ions.

Originally, the enzyme used for PCR is the large fragment of DNA polymerase derived from *E. coli* (klenow). The identification of an enzyme, Taq polymerase, that is stable at the high denaturing temperature makes it necessary to add the enzyme with each cycle and hence, has made it possible to automate the technique. Taq polymerase was obtained from the *Thermophillic bacterium*, *Thermus aquaticus*. The two primary oligonucleotides dictate the specificity of the reaction. The specificity can also be controlled by varying the reannealing and extension times, temperature and the enzyme concentration. Oligonucleotides with high G-C content generally require high reannealing and extension temperatures. These primers have different sequences that lie on opposite strand of the template.

The PCR process involves three cycles:

Denaturation: In this cycle, the double stranded DNA template get denatured due to high temperature

Annealing: After high temperature, it is cooled to a temperature that allows the primers to anneal to both the strands of the target DNA.

DNA synthesis or extension: In this cycle the DNA polymerase that is thermostable, also called as Taq DNA polymerase, extends the primers to form a full new segment of DNA.

These cycles are repeated several times to get a large amount of DNA.

The technique is based on the logarithmic amplification of the specific nucleic acid segment so that even minute trace amount of a particular infectious agent can become detectable. A diagrammatic presentation of PCR steps is give in Fig. 69.

Template

The basic template used for PCR is generally double stranded DNA. RNA or cDNA are also used which offer the advantage of fewer sequences being available for non-specific amplification. The viruses have their genome as DNA or RNA and the nucleic acid fragments are used as template for PCR reaction. The RNA from viruses having RNA genome is isolated and cDNA synthesis is done by the use of reverse transcriptase (RT) enzyme and this cDNA is used as template.

The methods for isolating DNA and RNA from symptomatic and healthy plant leaves are given below.

DNA EXTRACTION BY COMMERCIAL KIT (Qiagen, Germany): The protocol of the company may be followed. The major points are given below:

1. 150 mg leaves from infected plant are ground in liquid nitrogen in sterilized pestle and mortar.
2. The powder is transferred in to 1 ml eppendorf tube.
3. 400 ml of buffer AP_1 and 4 µl of RNase-A stock solution (100 mg/ml) to a maximum of 100 mg of ground plant tissue is added and vortex vigorously.
4. The mixture is incubated for 10 min at 65°C and is thoroughly mixed 2-3 times during incubation by inverting the tube.
5. 130 µl of buffer AP_2 is added to the lysate, mixed and incubate for 5 min on ice.
6. Centrifuge the tube for 5 min at 12000 rpm.
7. Apply the lysate to the QIA shredder spin column (Lilac) sitting in 2 ml collection tube and centrifuge for 2 min at 12000 rpm.
8. The flow through fraction is transferred to new tube without disturbing the cell debris pellet.
9. 1.5 volume of buffer is added to the cleared lysate and mix by pipetting.
10. 650 µl of mixture from above tube is transferred to DNase mini spin column sitting in 2 ml collection tube.
11. Centrifuge for 1 min at 6000 x g and flow through is discarded.
12. The above step is repeated with remaining sample and flow through was discarded.
13. The DNase column is placed in new 2 ml collection tube. Add 500 µl buffer AW to the DNeasy column and centrifuge for 1 min at 6000 x g.
14. 500 µl buffer AW is added to th DNeasy column and centrifuge for 2 min at maximum speed to dry the membrane.
15. DNeasy column is transferred to 1.5 ml micro-centrfuge tube and 50 µl of pre-heated at 65°C buffer AE was directly added into the

DNeasy membrane and is incubated for 5 min at room temperature and centrifuge for 1 min at 6000 rpm.

16. The above step was repeated.

Note: All the reagents etc. are supplied with the kit by the company.

DNA extraction by sodium sulphite method

1. 150 mg of leaves from infected plant is ground in liquid nitrogen.
2. 10 ml extraction buffer is heated to 65°C before adding to powdered tissue.
3. The mixture is taken into eppendorf.
4. 1 ml of hot extraction buffer is added.
5. The eppendorf containing powdered tissue in extraction buffer is kept at 95°C heating block for 10 min.
7. Do regular vortexing after every two min.
8. Keep on ice for 2 min.
9. Centrifuge the tube at 12000 rpm for 5 min.
10. Take out approximately 800 µl of supernatant and transfer it to new tube containing 5 µl of RNase.
11. Incubate at 37°C for 20 min.
12. Add 480 µl of isopropanol and mixed by gentle shaking.
13. Centrifuge the tube for 5 min at 12000 rpm.
14. Add 30 µl of sterile distilled water to the pellet.
15. To dissolve DNA pellet in distilled water, it is heated briefly at 50°C and flicking is done.
16. The DNA is precipitated with 30 µl of 3 M sodium acetate and $1/10^{th}$ volume of 95% of ethanol.
17. The tube is kept on ice for 10-20 min.
18. Centrifuge the tube at 12000 rpm for 5 min. and pour off ethyl alcohol.
19. 400 µl of 70% ethanol is added and vortexed.
20. Centrifuge for 3 min at 12000 rpm.
21. Pour off supernatant.
22. The pellet is dried for 40 min at 37°C.
23. Then the pellet was dissolved in 70-100 µl of double distilled water, preheated to 45-50°C.

Preparation of DNA extraction buffer for sodium sulphite method

1 M tris HCl (pH 8.0)	1.0 ml
3 M KCl	3.3 ml
5 mM EDTA	0.2 ml
Double distilled water	5.5 ml

Sodium sulphite	0.065 g
Total	10.0 ml

DNA extraction by CTAB method

1. 10 ml extraction buffer is prepared. (see Appendix at the end of this book) and heat at 65°C for 20 min.
2. 100 mg infected leaves are ground in liquid nitrogen.
3. 1 ml of heated buffer is added to the tube containing powdered leaf tissue.
4. Add 0.7 volume chloroform : isoamyl alcohol in a ratio of 24:1.
5. The tube is shaken by inverting it gently for 10-20 min.
6. Centrifuge at 12000 rpm at 4°C for 10 min.
7. The aqueous phase is removed from the tube.
8. 0.7 volume of isopropanol is added to the tube.
9. Centrifuge at 4°C for 7-10 min at 10,000 rpm.
10. The pellet is washed with 400 µl of 70% ethanol.
11. Dry the pellet at room temperature.
12. The pellet is dissolved in 30-50 µl of double distilled water with 2 µl RNase.

Note: For each method of DNA isolation, leaves from healthy plants must be processed same way to be used as control.

DNA extraction buffer for CTAB method

100 mM tris, pH 8.0	1.0 ml
2% CTAB	2.0 ml
1:4 M NaCl (5 M)	2.8 ml
B-mercapto-ethanol	200 µl
EDTA 0.5%	30.0 µl
Double distilled water	3.7 ml
Total	10.0 ml

Quantification of DNA

For quantification of isolated DNA, it is diluted in double distilled water. 5 µl DNA of each sample is diluted in 495 µl of double distilled water and record UV absorption at 260 nm and 280 nm. Readings were taken and average value of sample was calculated. The yield of DNA per µl was calculated with the following formulae.

$$\text{Ng of DNA per } \mu\text{l} = \frac{\text{UV absorption at 260} \times \text{dilution factor} \times 50}{1000} \times 1000$$

Extraction of RNA

Total RNA can be isolated from virus infected plants following manufacturer's instruction (QUAGEN)

1. Collect 100 mg virus infected leaves, wash them and dry.
2. Ground the leaves in liquid nitrogen.
3. Powdered samples are immediately soaked with 450 µl lysis buffer (RLT buffer).
4. Vortexed for 10 sec and incubate at 56°C for 3 min in water bath.
5. Clear lysate is then passed through another spin column and centrifuge at 8000 g for 2 min at room temperature.
6. The flow-through is transferred to a new eppendorf tube and half volume (225 µl) of chilled ethanol (absolute) is added to clear lysate and mixed by pipetting.
7. Passed through another spin column (pink column) and centrifuge at 8000 g for 1 min at room temperature.
8. Flow-through is discarded and 700 µl of RW 1 buffer was added to the column.
9. Centrifuge for 1 min at same speed.
10. Again flow-through and collection tubes are discarded. 500 µl RPE buffer is added to the same column and centrifuge for 1 min.
11. The same is repeated and centrifuge for 2 min. Finally RNA is eluted with 30 µl sterile RNase free water by centrifuging at 8000 g for 1 min and is collected in an eppendorf.
12. This preparation was used as a template for reverse transcription and RT-PCR.
13. Sap extracted from healthy leaves and processed similarly is used as control.

A simple protocol to prepare nucleic acid extracts suitable for woody as well as herbaceous plants has been developed utilizing 0.65 to 0.7% of sodium sulphite. The procedure consists of adding 0.65% (w/v) sodium sulphite to Tris-HCl extraction buffer containing DNase-1 and $MgCl_2$. Six drops (150-200 µl) of leaf extract are added to the buffer and same mixture is incubated for 10 min at 37°C. The nucleic acids are extracted with water saturated phenol and re-extracted with phenol : chloroform : isoamyl alcohol (v:v:v, 25:24:1), followed by precipitation of RNA with isopropanol. This protocol is better than commercially available kits for the extraction of RNA and also applicable to various plant tissues *e.g.* leaves, tuber, and sprouts.

Extraction of RNA can also be obtained by following protocol:

1. Take 100 mg leaf tissue, wash and dry.
2. Ground to powder in liquid nitrogen in a microfuge tube and extract

in 300 µl of 0.1 M Tris-HCl, pH 8.0, 2 mM EDTA and 2% sodium dodecyl sulfate (SDS).

3. 300 µl of phenol chloroform (1:1) was added to the reaction tube.
4. Vortexed and heated at 70°C for 5 min.
5. Centrifuge for 5 min.
6. Pass the supernatant through a 1 ml Sephadex G-50 column.
7. Store the elute in liquid nitrogen
8. A one tube reaction for reverse transcription of RNA to cDNA and its subsequent application will be described later in this chapter.

Measurement of RNA

1. Turn the ultraviolet (UV) spectrophotometer and set the wavelength to 260 and 280 nm according to instructions.
2. Set up a reference using blank solution. Depending on the RNA sample, add 1 ml of TE buffer (pH 8.0) or double distilled water to a clean cuvette (one half to three quarters full), depending on whether one or two cells are available in the spectrophotometer. Insert the cuvette(s) into cuvette holder(s) in the sample compartment with the optical (clear) side of the cuvette(s) facing the light path. Close the sample compartment cover and adjust the number to 0.000 (either manually or by pressing the 'Auto Zero' button, depending on the specific spectrometer.
3. Check the mismatching of quvettes by filling cuvette(s) with blank solution and recording the absorbance (positive or negative number). Make appropriate additive (for negative number) or subtractive (for positive number) corrections to future RNA samples. In most cases, however, it is optional.
4. In a clean cuvette approximately dilute the RNA sample to be measured in blank solution to a total volume of 1 ml. For example for 2,5,10,15,20 or 25 µl of RNA sample, the blank solution should be 998, 995, 990, 985, 980, or 975 µl respectively. Wrap the cuvette in a piece of parafilm and mix well by inverting the cuvette two to three times.
5. Insert the sample cuvette and read the absorbance of the sample cuvette against the blank cuvette according to the spectrophotometer instructions. If only one cell is available, remove the reference cuvette after adjusting the number to 0.000 and insert the sample cuvette.
6. After measurement RNA samples at 260 and 280 nm respectively, calculate the ratio of A260/A280 reading numbers and concentration for each RNA sample. Some spectrophotometers (*e.g.* UV 160U, Shimatzu) can automatically calculate, display,

and point out the ratio and concentration of each RNA sample. A pure RNA should have a ratio of 1.85 to 2.0 of A260/A280 reading numbers. The concentration of RNA sample can be manually calculated as follows:

Total RNA (µg/µl) = A 260 reading number x 40 µg/ml x dilution factor

For example, if 25 µl of RNA is diluted to 1000 µl for measuring and its A 260 reading number is 0.4000, then

Total RNA(µg/ µl) = A 260 reading number x 40 µg/ml x dilution factor
= 0.5000 x 40 µg/ml x 1000/25
= 800 µg/ml
= 0.8 µg/µl

Selection of primers

1. Primers are short segments of nucleic acid (often 20-30 nucleotides length) which attach (by specific bindings of their bases with those of corresponding bases in the segment which is being targeted for diagnosis) to the beginning and the end of the specific target segment of nucleic acid.
2. By using the knowledge of sequence of the specific nucleic acid segment, primers can be designed which would attach either end of the target sequence.
3. After attachment to either end of the segment, the primers "kick start" the polymerization process of new nucleic acid.
4. Primers can be synthesized from "Genset Oligos: a division of Genset". Two primers, one forward and one reverse are used for amplification of the coat protein gene.

cDNA Synthesis

The RNA isolated from infected leaves by the methods described above is used as template to perform the following reaction for cDNA synthesis.

Template RNA	10 µl
RT buffer 10X	2 µl
dNTPs 10 mM	2 µl
Reverse primer (100 ng/µl)	2 µl
Reverse transcriptase (M-Mul V, 20 units/µl)	1 µl
RNAs free water	2 µl
RNAs inhibitor	1 µl
	20 µl

Template RNA incubated at 70°C for 5-10 min, chilled on ice (3-5 min) and other reactives are added to the tube. The tube is chilled on ice and given a quick spin to bring down the contents. For the first strand cDNA synthesis, the reaction mix is incubated at 42°C for 45 min.

PCR Amplification

For PCR amplification to an aliquot of cDNA, the following reactives are added.

Sterile distilled water (SDW)	26.50 µl
10 x PCR buffer	5.00 µl
2 mM dNTPs (deoxynucleotide Triphosphates)	1.00 µl
25 mM $MgCl_2$	2.00 µl
Forward primer (100 ng/µl)	2.00 µl
Reverse primer (100 ng/µl)	3.00 µl
25 unit Taq Polymerase	0.50 µl
RNA template (cDNA)	10.00 µl
	50.00 µl

The reaction mixture is kept in a Thermal cycler (PCR matchine) for amplification of the target molecules using the conditions given in Table 4.

Table 4. PCR amplification schedule

Cycle	Denaturing	Annealing	Polymerization
1st cycle	94°C, 5 min	—	—
2nd cycle (of 30 cycles)	94°C, 1 min	45°C, 1 min	72°C, 3 min
Last cycle	94°C, 1 min	45°C, 1 min	72°C, 3 min

Agarose Gel Electrophoresis

The PCR amplified product is resolved on a 1% agarose gel as follows. 0.4 mg of agarose is added to 40 ml 1 x TAE (prepared in sterile water), boil and cool up to 50°C. 1 µl of ethidium bromide (10 µg/100 ml stock) is mixed with 6x loading dye (orange G or Bromophenol blue) and was added to agarose. This agarose mix is loaded to electrophoresis unit. Electrophoresis is run at 50-60 volts for 1-1.5 h. Results (nucleic acid bands) are evaluated on a UV transilluminator light (Fig. 70).

However, these primers amplify only fragments from the coat protein of the virus. Another set of primers will be required to design for amplification of the putative coat protein gene. They should contain the putative ATG start codon and an adapter arm with *BamH 1* restriction site and a putative stop codon and an adapter arm with *EcoR 1* restriction site.

Using different sets of primers, the annealing temperature, condition of amplification and final extension may vary. This may result amplification of longer fragment of coat protein gene (Fig. 71).

Summary of PCR test

A. **Selection of primers:**

It has already been described above.

B. **Extraction of DNA from test samples:**

1. Standard technique such as phenol extraction is used to remove the DNA from test samples.
2. For detection of RNA viruses, a reverse transcriptase step to convert RNA to complementary DNA (cDNA) is added as the standard PCR test can only be done on DNA.

C. **Denaturation of DNA:**

1. Primers will only attach to and subsequently elongate the developing nucleic acid chain on a template of single stranded DNA (ssDNA).
2. Thus, if the target sequence is double stranded-either because it belongs to a dsDNA virus or the segment under investigation is part of a bacterial or cellular chromosome, the strands need to be split apart from each other to produce ssDNA. This is achieved by heating the material to 90-95°C – a process termed as denaturation.

D. **Annealing of primers:**

The primars are then attached to the ends of the segments to be amplified – a process called annealing – this takes place at a temperature range of 37-50°C.

E. **Polymerization:**

1. After the primers are attached they "kick-start" the polymerization which then elongates the developing nucleic acid chain between them.
2. The enzyme Taq polymerase is from hot springs and it is therefore, active at a very high temperature (72°C).
3. Bases are successively added on to the developing new nucleic acid chain according to the sequence of the segment, which is being detected, producing a new chain consisting of a complementary sequence to that of the target segment.

F. **Amplification:**

1. The new dsDNA-(one strand of the old template and one strand the mixture at 95°C *i.e.* denaturation, and then the cycles from step No. C-E are repeated.

2. The cyclical variation of temperature of the reaction mixture, *i.e.* 95°C (denaturation), 45°C (annealing) and 72°C (polymerization) is carried out by an automated thermal cycler.
3. The cycle is repeated a number of times – each cycle results in a logarithmic increase in the amount of DNA which is amplified.
4. Thus, within 20 cycles a million fold amplification of the starting amount of nucleic acid can be achieved.

G. Detection of target DNA sequence:

Having achieved amplification from a minute amount of starting target sequence nucleic acid to reach significant amounts of DNA – this can now be readily demonstrated by conventional DNA detection techniques such as nucleic acid hybridization as described earlier or by agarose gel electrophoresis.

Multiplex PCR

It is quite common that a plant may be infected with more than one virus/viroid (Fig. 72). To test each virus/viroid individually is time taking and expensive, it was therefore realized to develop techniques to test several viruses/viroids in a single PCR reaction. Multiplex PCR was therefore evolved for the first time in 1988 (Chamberlain *et al.*, 1988) and is now widely used for detection of pathogens in humans and for viruses in plants. This technique is used when more than one target is to be amplified simultaneously in a single reaction. It results in a considerable saving of time, reagents and labour in sample handling. Many crops and trees harbor multiple viruses or viroids, hence multiplex PCR is found to be very useful in diagnosis of these pathogens simultaneously.

Multiplex PCR requires re-optimization of reagent concentrations for example an increase requirement of Mg^{2+}, dNTPS, polymerase, primer concentration and proportion of each individual primer pairs. The following

Table 5. Comparison of PCR mix in 25 µl reaction

	Uni to multiplex			
Reagent	**Uniplex**	**Duplex**	**Tetraplex**	**Pentaplex**
PCR buffer	1x	1x	1x	1x
$MgCl_2$ (mM)	1.5	2	3.5	4
dNTPs (mM)	0.2	0.2	0.4	0.5
Antisense primer (ng/ µl)	2	4	6	8
Sense primer (ng/µl)	2	2	2	2
Taq (U)	0.625	0.625	0.625	0.625

1 x buffer = 10mM tris-HCl, pH 8.3 and 50 mM KCl.

dNTPS = dATP, dCTP, dTTP and dGTP

Table 5 shows the typical values of the increased amounts of various reagents in multiplex RT-PCR for the detection of potato virus strains (R.P. Singh, 2009).

Nucleic acid extraction

Nucleic acid extract can be prepared according to the manufacturer's instructions, using commercial RNeasy kit (Qiagen). If the kit is not available then prepare RNAs by the sodium sulphite procedure. The procedure in brief is:

1. Homogenize 100 mg of leaf tissue in an extraction buffer (0.1 M Tris-HCl, pH 7.4, 2.5 mM $MgCl_2$, and 6 units of DNase 1) with a mortal and pestle.
2. Incubate the homogenate at 37°C for 10 min.
3. Extract the RNA with phenol : chloroform : isoamyl alcohol (V:V:V; 25:24:1) using equal volume of sap and organic solvents.
4. Precipitate the RNA with equal volume of isopropanol containing 0.1 volume of 3 M sodium acetate, pH 5.2.
5. Keep at -20°C overnight
6. Collect the RNA by centrifugation at 12000 x g for 15 min at 4°C
7. Wash with 70% ethanol, dry at 37°C for 20-30 min.
8. Finally dissolve in 100 µl of sterile distilled water.

Design of primers for multiplex PCR

Design and selection of primers suitable for multiplex RT-PCR requires consideration of primer specificity; primer Tm compatibility; and more importantly, the size of amplicons. In uniplex PCR, in which only one pair of primer is used, primer specificity and size of amplicon are the primary factor to be considered. Generally, within a certain fragment length *e.g.* 100-1000 bp, the shorter the targeted fragment the more efficient the amplification (Fig. 76). In contrast, all targeted fragments in the multiplex-RT-PCR should be in the proper fragment length range to be efficiently amplified during PCR and should also be sufficiently different in size to permit efficient separation in the agarose gel.

Multiplex PCR has now been used to detect viruses in several crops such as potato, citrus, banana etc (Figs. 73, 74 and 75).

Reverse transcription

cDNA of potato leaf roll virus (PLRV), potato virus X (PVX), and Potato virus Y (PVY) were prepared using oligo (dT) common primer. For the separation of PVY strains, the cDNAs were prepared using 50 ng of antisense primers as given in the Table below. cDNA was synthesized using reverse transcription system (Promega). The reverse transcription mixture was as given below:

1 x buffer (10 mM tris-HCl, 50 mM KCL, 0.1% Triton X-100)

5 mM $MgCl_2$

1 mM of deoxyribonucleotide (dNTPs)

20 units of RNase

0.5 ng Random primers

9 µl viral RNA (which was first denatured at 65°C and chilled on ice)

15 U of AMV RT

These were added to make 20 µl total reaction volume. The mixture was incubated at 42°C for 1 h. Thereafter, AMV RT was inactivated upon incubation at 95°C for 2 min and the total volume was made up to 100 µl by adding 80 µl of nuclease free water.

PCR amplification:

The antisense and sense primers used are mentioned in the above Table 6. The PCR conditions varied depending on the number of intended viruses to be detected and the number of primers involved. For uniplex PCR, 5 µl of the cDNA mixture and the following reagents were added: 2.5 µl of PCR 10x buffer [100 mM tris-HCl, 500 mM KCl, 0.01% (w/v) gelatin], 1.5 µl of $MgCl_2$ (25 mM), 2.5 µl of dNTPs mix (2 mM each), 50 ng (1 µl) each of specific sense and antisense primers, 0.5 µl AmpliTaq (5 unit/µl, Perkin-Elmer, Applied Biosystems) and the reaction volume was made to 25 µl with nuclease free water. Cycling conditions were optimized to an initial denaturation at 94°C for 5 min and 30 subsequent cycles of amplification with 60 sec of denaturation at 92°C, 1 min of annealing at 60°C and 1 min of extension with 72°C followed by one cycle of final extension for 10 min at 72°C. In case of duplex and multiplex PCR the $MgCl_2$ concentration was raised to 2.5 µl of $MgCl_2$ (25 mM). 10 µl of PCR product were electrophoresed in 1.5% agarose gel containing 0.5 µg/ml of ethidium bromide and photographed under UV light. The results of these tests are shown in Fig. 76.

Table 6. The primer pairs used for various potato viruses and PVY strains in RT-PCR

Virus		Sequence	Fragment (bp)
PLRV	4S	5'-CGCGCTAACAGAGTTCAGCC-3'	336
	5A	5'-GCAATGGGGGTCCAACTCAT-3'	
	24F2	5'-CACTGTGCTCCCTTAAATCC-3'	613
	24A	5'-CTCACCTCCATATCATCCTCC-3'	
PVX	10S	5'-TAGCACAACACAGGCCACAG-3'	562
	10A	5'-GGCAGCATTCATTTGAGCTTC-3'	
PVY^o strain	1S	5'-ACGTCCAAAATGAGAATGCC-3'	480
	1A	5'-TGGTGTTCGTGATGTGACCT-3'	
	S7	5'-GACAGTTGGACTTTTGCAACG-3'	281
	A	5'-CATTTGTGCCCAATTGCC-3'	
PVY^n strain	S6	5'-GGTGAAGCTAATCATGTCAAC-3'	443
	A	5'-CATTTGTGCCCAATTGCC-3'	

Multiplex PCR is not only used for the detection of viruses but also the fastidious bacterium, phytoplasma and other pathogens.

Optimization of the multiplex PCR detection of viruses

In RT-PCR, since the quality of cDNA determines the efficiency of amplification to a great extent, the standardization of the cDNA preparation minimizes the concentration problem often encountered with antisense primers in developing multiplex RT-PCR. To minimize the number of antisense primers for cDNA preparation, a degenerate primer which share a high degree of sequence identity at the 3'-end to multiple templates may be used. Alternatively, use the commercially available random oligonucleotides (hexanucleotides containing 4 different nucleotide combinations) as primers for cDNA preparations of all types of RNAs. Another approach is the use of random primers generated during RNA isolation protocol in which DNase 1 is used to digest and eliminate host genomic DNA to prime the reverse transcriptase

Multiplex PCR can differentiate strains and sub-strains of a single virus by conventional multiplex RT-PCR, hemi-nested RT-PCR (such as P1 gene has been targeted for strain differentiation using RT-PCR) and 'competitive' multiplex RT-PCR (by using a common antisense primer for all the strains of the targeted virus and by designing the individual sense primers utilizing the small differences available in the nucleotide stretches in the genome of strain, a multiplex RT-PCR can be performed).

In conclusion, multiplex RT-PCR is a valuable tool for a simultaneous detection of different viruses, viroids and fastidious bacteria; and for simultaneous differentiation of various strains of the same virus. Primer design and selection are the critical steps for successful application of multiplex RT-PCR.

DIFFERENT TYPES OF PCR

RT-PCR : Most plant viruses have RNA as their genome. Since dsDNA is needed for PCR, the isolated RNA will be converted to cDNA through reverse transcriptase (RT). PCR can be performed only after obtaining cDNA of the RNA strand using a first step of RT, hence it is called as RT-PCR.

Immunocapture-RT-PCR

This technique is now used for large scale detection of plant viruses. The test can be performed in separate tubes or ELISA plate or in the eppendorf tubes in which the PCR reaction is to perform. Ideally, the test can be performed in plates that have already been used for complete ELISA assay. The test can detect the virus in samples which failed to be detected in ELISA. Therefore, in this technique virus particles are first trapped by specific antiserum and then PCR is performed (normally if ELISA results are negative).

Multiplex RT-PCR

In this technique several viruses and viroids can be amplified simultaneously using a cocktail of primer pairs. The method has already been described in this chapter.

Real Time PCR

An advancement in multiplex PCR is the development of Real Time PCR. It is a quantitative procedure, which helps to detect accumulation of PCR products during the PCR reaction. Real time quantitative PCR instrument (TaqmanR) is used in which ABI prism 7900 continuously measures PCR products accumulation using a dual labeled fluorogenic oligonucleotide probe called a Taqman probe. This probe is labeled with two different fluorescent dyes-the 5' terminus reporter dye and the 3' terminus quenching dye. The sequence of oligonucleotide probe is homologous to an internal target sequence present in the PCR amplicon. When the probe is intact energy transfer occurs between the flourophores and the fluorescent emission is quenched. During the extension phase of PCR the probe is cleaved by 5' nuclease activity of Taq polymerase. Therefore, the reporter is no longer in proximity with the quencher and the increase in emission intensity is measured. This technology uses different fluorescent labels attached to nested primers and can detect and quantify in to single reaction up to three pathogens in a time without having to run an agarose gel. Real time PCR has been marketed by Applied Biosystems.

Advantages of PCR

1. The technique is highly sensitive.
2. It can detect one molecule of nucleic acid within 100,000 cells.
3. Because of selective amplification it can pick up a minute amount of target nucleic acid within an ocean of extraneous cellular or other non-specific DNA.
4. PCR can detect infections at an early stage.
5. The technique is very useful for detection of non-replicating viruses.
6. It is more rapid technique than other diagnostic techniques.
7. The detection results can be obtained within 24-48 h.
8. Several viruses or viroids can be detected simultaneously in single multiplex RT-PCR.

Disadvantages of PCR

1. Because of the higher magnifying ability of the test, there is always a risk of contamination of the system which will result in false positive reactions.
2. The reagents used in the test are very expensive.

DOUBLE STRANDED RNA (dsRNA) TECHNOLOGY FOR IDENTIFICATION OF VARIATION IN PATHOGENICITY OF A VIRUS

This technology was considered to be very useful at one time specially for detection of strains in citrus tristeza virus (CTV). It was observed that when CTV is passed through its different hosts, it showed a different dsRNA pattern (Fig. 77).

For dsRNA analysis, dsRNA from young bark (for CTV) was analysed by phenol extraction of nucleic acids, purification by CF-11 cellulose column chromatography and separation by polyacrylamide gel electrophoresis. dsRNAs were observed by staining with ethidium bromide.

dsRNA profiles induced by a virus when passed through different hosts shows similar band, corresponding to the full length replicative form of the genome, but shows differences in the sub-genomic bands. Variations in pathogenicity and dsRNA profile have been observed when some CTV isolates were passed through different hosts. The technology established that many CTV isolates are probably a mixture of strains. Similar situation may also exist with other viruses. Composition and/or relative concentrations of individual strains in the mixture may change according to the host giving rise to the variation in symptom intensity and/or dsRNA pattern.

It has been described that the virus or virus-like pathogens will be detected by several methods right from host infectivity to PCR. The sensitivity of each diagnostic test is given in Table 6a.

Table 6a. Approximate sensitivity of virus detection

Techniques	Minimum detectable virus concentration
Host infectivity	100 ng
Electron microscopy	100 ng
Tube precipitation	500 ng
Gel diffusion	1000 ng
Latex test	5-20 ng
ELISA	1-10 ng
ISEM	1-10 ng
Western blot	1-10 ng
NAH	<1 pg
PCR	10 fg/up to attogram RNA

PLATE-I

Fig. 1. *Chenopodium amaranticolor* leaf showing necrotic lesions upon inoculation with turnip mosaic virus.

Fig. 2. *C. quinoa* leaf showing chlorotic lesions developed due to inoculation by turnip mosaic virus.

Fig. 3. Symptoms of dwarfing of large cardamom plant due to virus infection causing foorkey disease.

Fig. 4. Cucumber leaf showing mosaic symptoms by natural infection with cucumber mosaic virus.

Fig. 5. French bean leaf showing mottling symptoms when inoculated by Indian citrus ringspot virus (ICRSV).

Fig. 6. Leaves on an apple tree infected with apple mosaic virus showing yellow mosaic symptoms

PLATE-II

Fig. 7. Flecking on leaves due to infection of ICRSV (left) and citrus tristeza virus (CTV) (right).

Fig. 8. Yellowing of leaves due to infection by mungbean yellow mosaic virus.

Fig. 9. Greening of veins on a cucumber leaf due to a potyvirus infection.

Fig. 10. Yellowing of veins in bhendi (okra) infected with bhendi yellow vein mosaic virus.

Fig. 11. Citrus leaves infected with an isolate of citrus vein clearing virus showing vein clearing symptoms.

Fig. 12. Line pattern symptoms on plum leaves due to infection of apple mosaic virus

Fig. 13. Vein banding symptoms on apple leaves infected with an isolate of apple mosaic virus.

PLATE-III

Fig. 14. Leaves of kinnow mandarin showing ringspot symptoms when inoculated with ICRSV.

Fig. 15. Stem necrosis of a sunflower plant infected with an ilarvirus. Left is a healthy plant.

Fig. 16. Malformation and epinasty of leaves in Etrog citron infected with exocortis viroid. Right is a healthy plant

Fig. 17. Leaf roll symptoms on a potato plant infected with potato leaf roll virus.

Fig. 18. Tomato leaf curl virus infected tomato plant showing leaf curl symptoms.

Fig. 19. Leaf reduction and distortion of leaves in papaya plant infected with papaya ringspot virus.

Fig. 20. Brinjal plant infected with a phytoplasma showing leaf rosetting.

Fig. 21. Pits in the phloem of wood due to CTV infection.

PLATE-IV

Fig. 22. Necrosis of tissues in French bean leaf infected with ICRSV

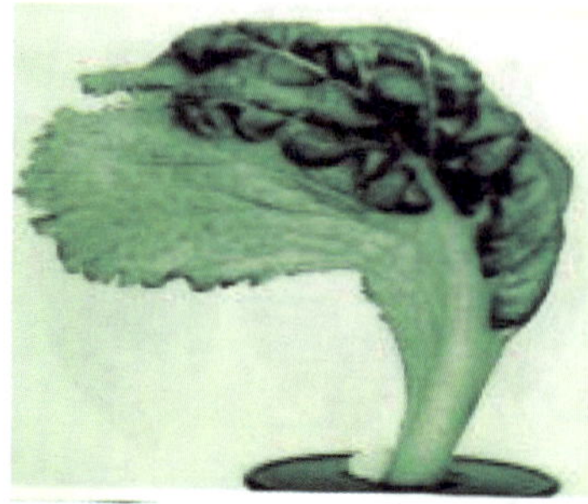

Fig. 23. Blisters on a leaf of *Brassica juncea* formed due to infection of turnip mosaic virus.

Fig. 24. Twigs of Kagzi lime plant showing tumors over them due to infection of citrus tumor virus.

Fig. 25. Apple fruits showing rough skin symptoms due to infection of a virus.

Fig. 26. An onion plant infected with onion yellow dwarf virus showing poor bud formation (left) as compared to healthy plant (right).

Fig. 27. Lime plants showing witch's broom symptoms when inoculated with lime witch's broom phytoplasma.

Fig. 28. Apple plant showing bending of branch due to flexibility occurred due to apple rubberywood phytoplasma infection.

PLATE-V

Fig. 29. Phyllody in a sesame plant caused by sesamum phyllody phytoplasma. Left is a healthy plant.

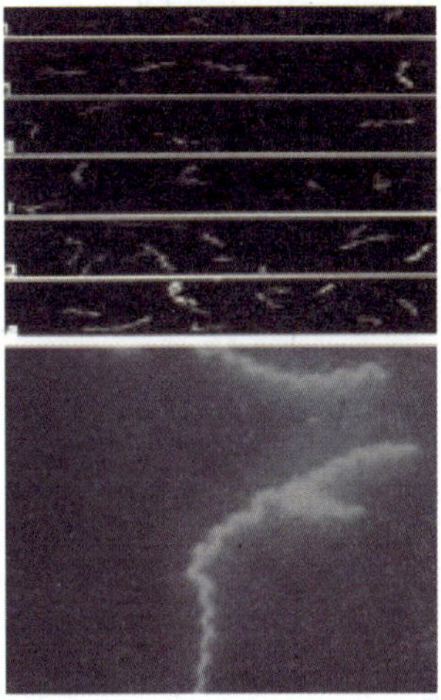

Fig. 30. *Spiroplasma citri* – causal organism of citrus stubborn disease in culture. Note the spiral form.

Fig. 31. Spindle shaped potato tubers due to infection of potato spindle tuber viroid. Right one is a healthy potato.

Fig. 32. Yellow corky vein symptoms in leaves of Kagzi lime infected with citrus exocortis and hop stunt viroids..

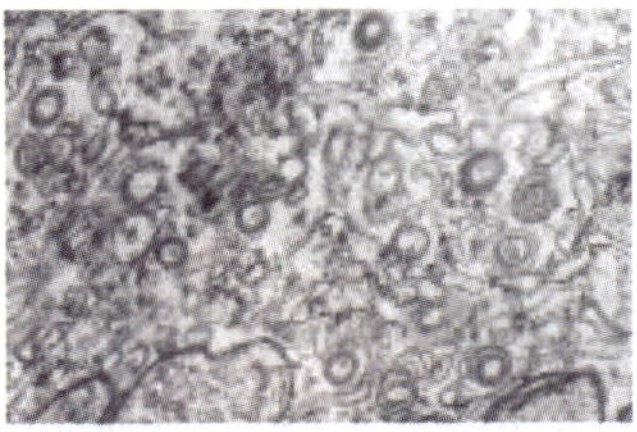

Fig. 33. A thin section of leaf tissues infected with a potyvirus showing pin wheel inclusions.

PLATE-VI

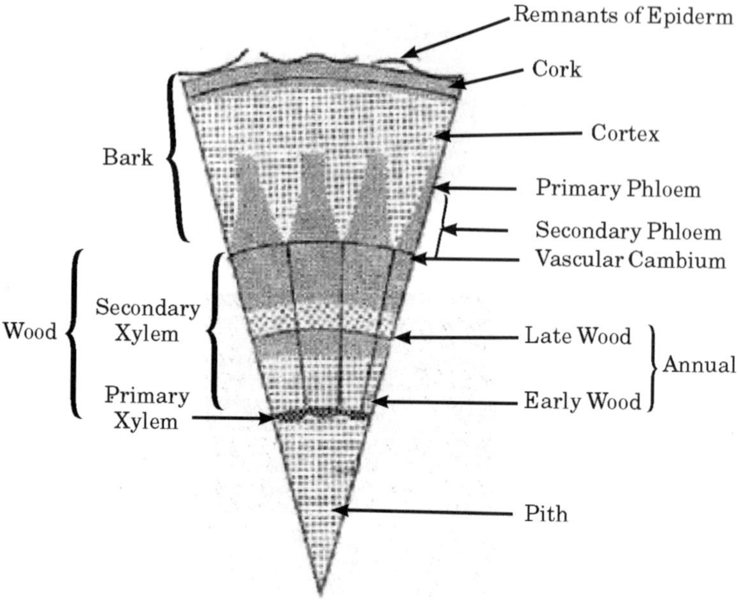

Fig. 34. Cross section of a woody plant showing cambium and other parts. Union of cambium between a healthy and infected plant facilitates virus transmission.

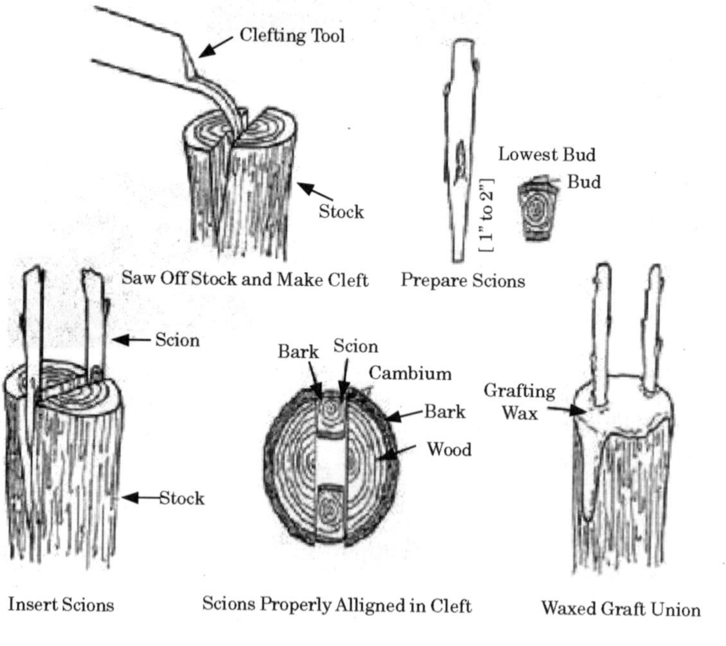

Fig. 35. Various stages of cleft graft are shown for successful grafting between healthy rootstock and infected scions for successful virus transmission.

PLATE-VII

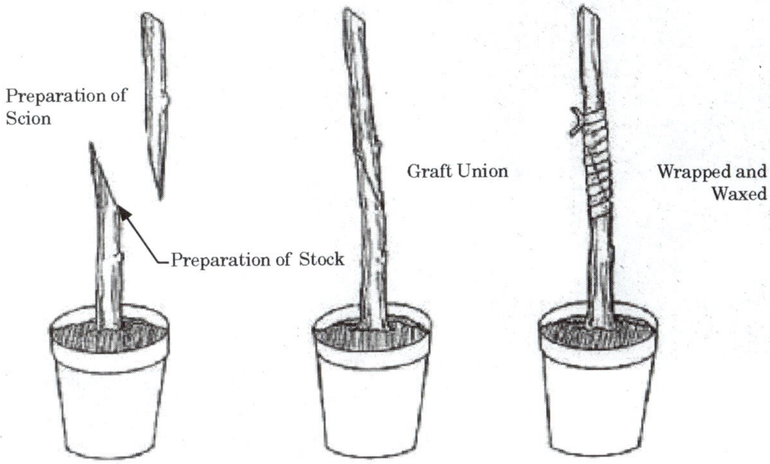

Fig. 36. Various stages of splice graft are shown.

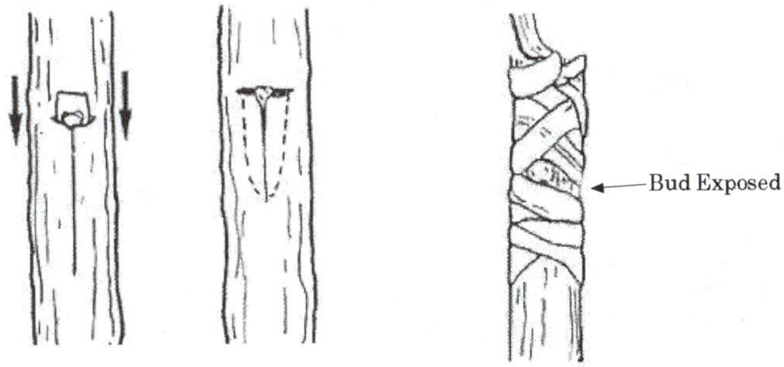

Fig. 37. Various stages of T-budding normally used for virus transmission of woody plants.

Fig. 38. Dodder, a parasitic plant is used for virus and phytoplasma transmission by way of making a bridge between infected and healthy plants.

Fig. 39. Banana aphids feeding on banana leaf infected with banana bunchy top virus for acquisition of virus during a transmission test.

PLATE-VIII

Fig. 40. Upper line **a.** Cages used for white fly transmission **b.** Fixing of whitefly cage to plant for transmission studies **c.** Cage used for feeding of leafhoppers on infected source. Second & third line: Different types of insect vectors of plant viruses. Second line: Aphids and mealybugs (centre), Third line: *Citrus psylla*, whitefly, thrips (upper), leafhoppers (lower).

Fig. 41. Wooden cages used for maintaining insect vectors (leafhoppers, psyllids and whitefly etc.) in an insectory.

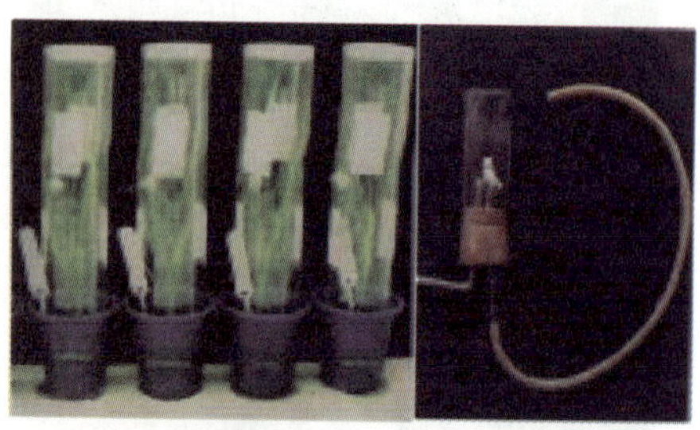

Fig. 42. Plastic cages used for leafhopper transmission studies. Right is an insect catcher used for collecting insects for transmission tests.

PLATE-IX

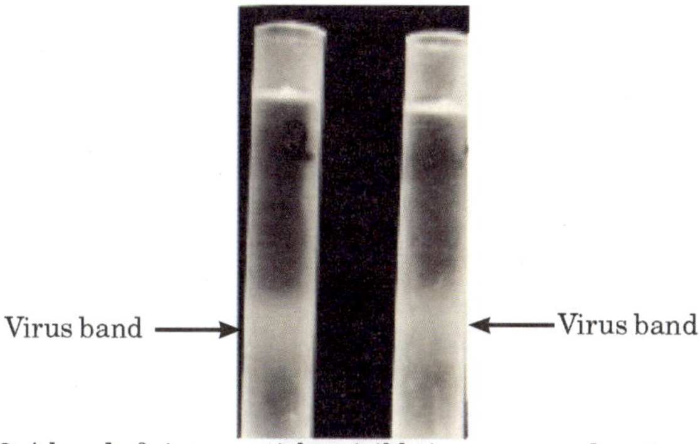

Fig. 43. A band of virus particles visible in a sucrose density gradient.

Fig. 44. Electron microscope at the Advanced Centre for Plant Virology at IARI, New Delhi

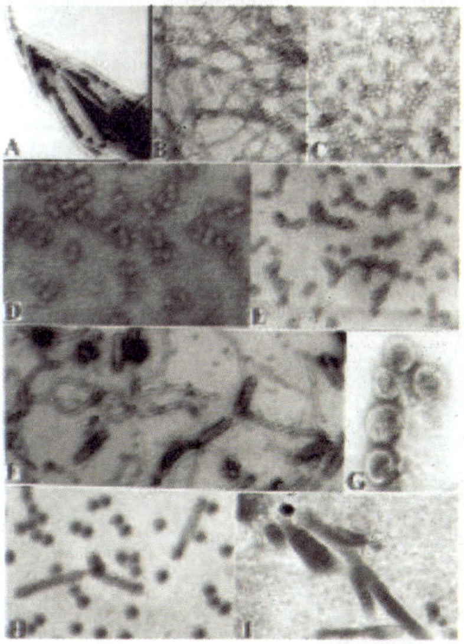

Fig. 45. The electron micrograph shows different morphology of virus particles as: **First line:** Tobacco mosaic virus, potato virus Y, cucumber mosaic virus **Second line:** Geminivirus, alfalfa mosaic virus; **Third line:** Mixed infection of citrus mosaic badnavirus, CTV, ICRSV; and **Fourth line:** Rice tungroviruses and citrus greening bacteria.

PLATE-X

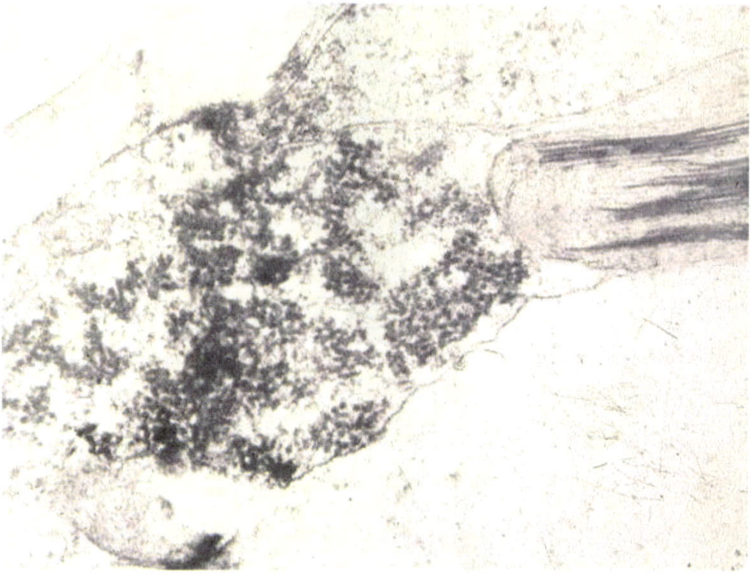

Fig. 46. Ultrathin section of tissues infected with citrus yellow mosaic virus showing bacilliform particles in cytoplasm of cell.

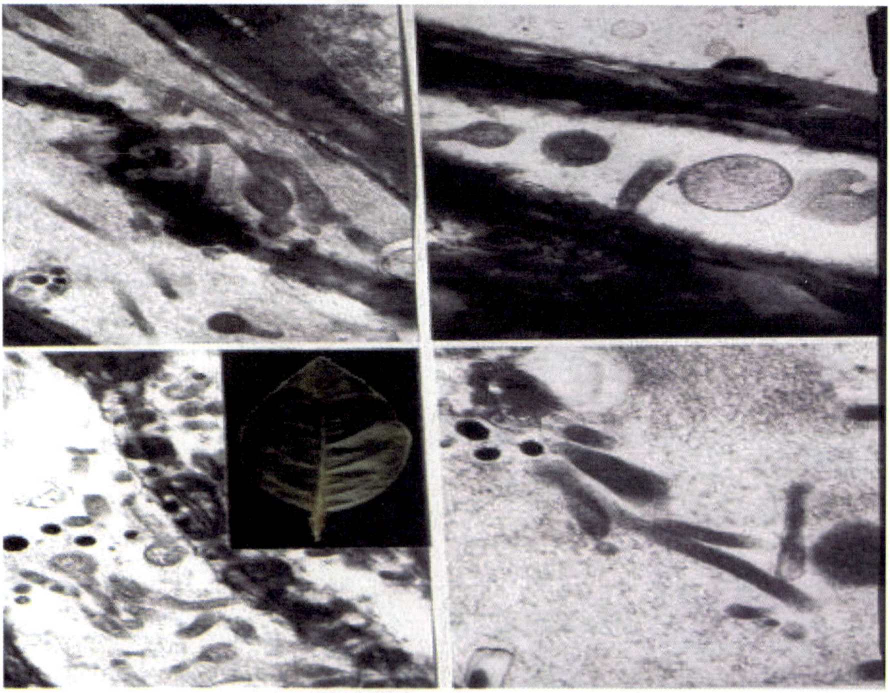

Fig. 47. Ultrathin sections of leaf midrib of greening infected citrus plant showing pleomorphic bodies of the bacterium in sieve tubes of phloem cells.

PLATE-XI

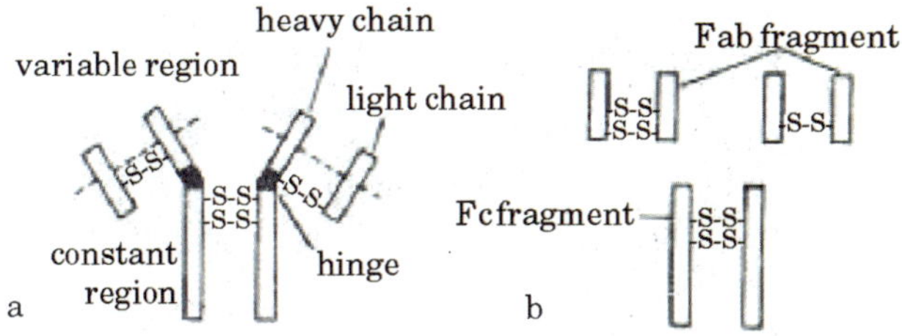

Fig. 48. Structure of antibody (a) the antibody molecule consists of four chains of protein : two light chains and two heavy chains connected by disulfide bond. The heavy chain bond at hinge point. Variable region is that portion of the molecule where the amino acids vary among antibodies. (b) with papain enzyme treatment, cleavage occurs at the hinge point resulting three fragments, two Fab fragments and one Fc fragment.

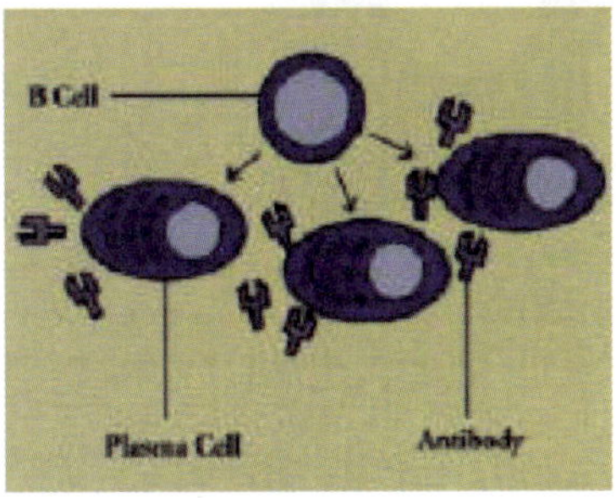

Fig. 49. Plasma cells of beta cell producing antibodies.

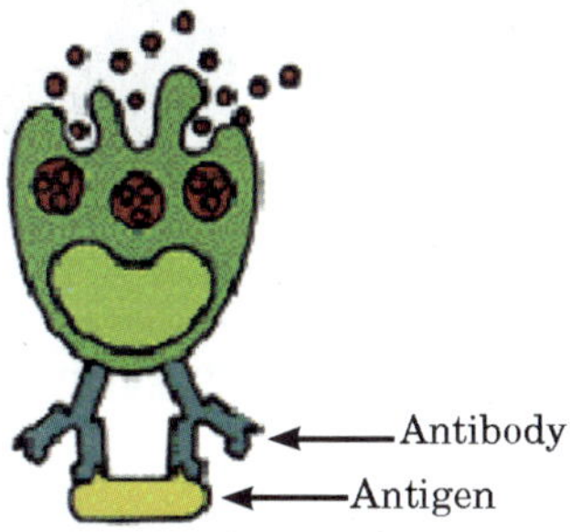

Fig. 50. Antigen-antibody in a reaction showing lock and key arrangement.

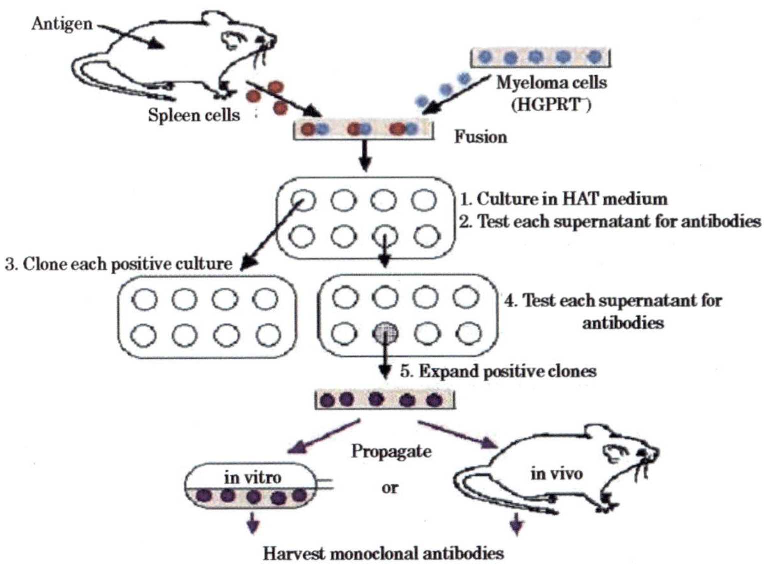

Fig. 51. Diagramatic presentation of the protocol for developing monoclonal antibodies (MAbs).

PLATE-XII

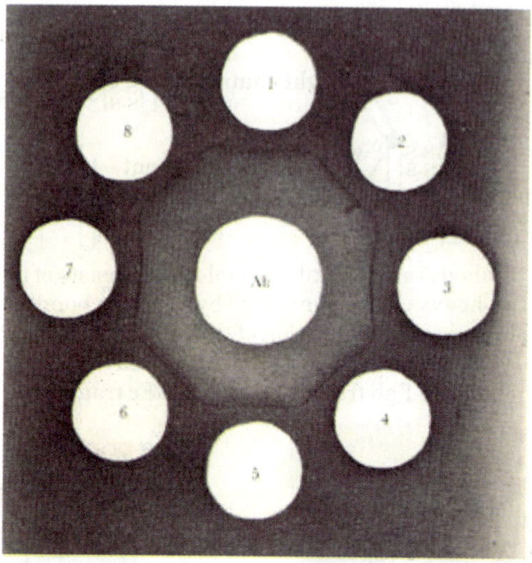

Fig. 52. *Gel diffusion test.* The antiserum, AB is placed in the centre well and the homologous antigen in the outer wells (1-8). A continuous precipitation line is developed where antibodies-antigen meets. (*Source*: Agricultural Canada Research Station, Vancouver)

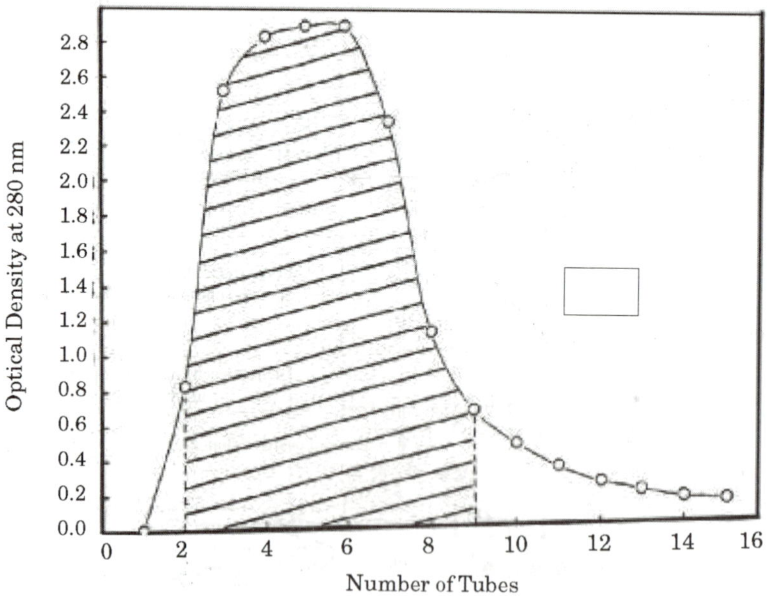

Fig. 53. Elution curve for immunoglobulin from DEAE Column

PLATE-XIII

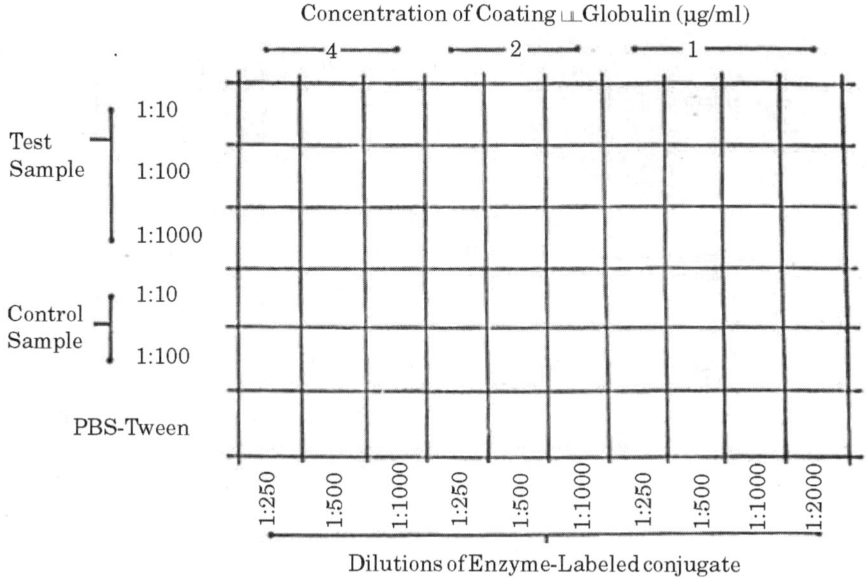

Fig. 54. Scheme for determining optimum concentration of coating and enzyme labeled γ–globulins.

Coating the well of ELISA plate with immunoglobulin.

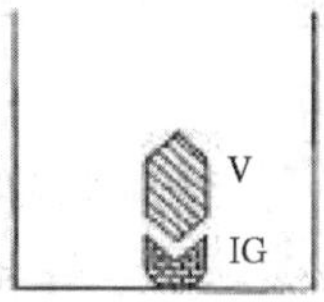

Add antigen which will attach with the antibody

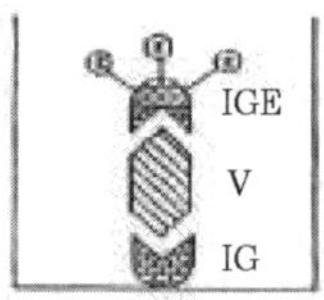

Add alkaline phosphatase (AP) conjugated antibody which will also attach to the antigen

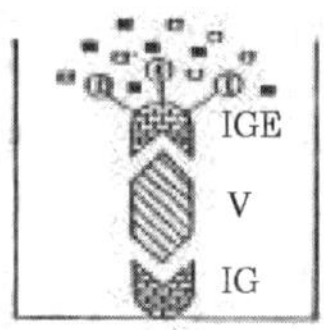

Add substrate (para-nitrophenyl phosphate) which will be hydrolysed by AP and yellow colour will develop (Fig. 56)

Fig. 55. Diagramatic presentation of ELISA

PLATE-XIV

Fig. 56. ELISA plate showing yellow colour in some wells which develops only with infected samples.

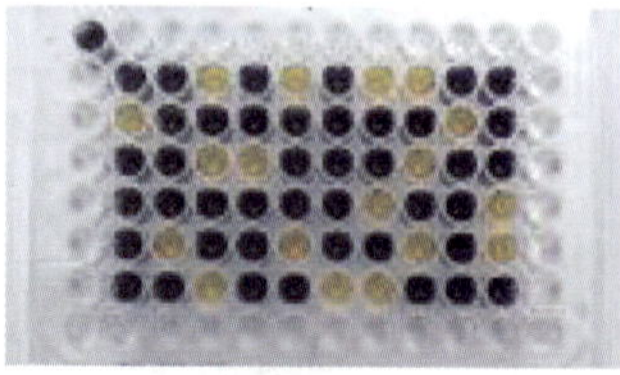

Fig. 57. Detection of PVY using penicillinase based ELISA kit (note the disappearance of black colour in case of virus positive wells).

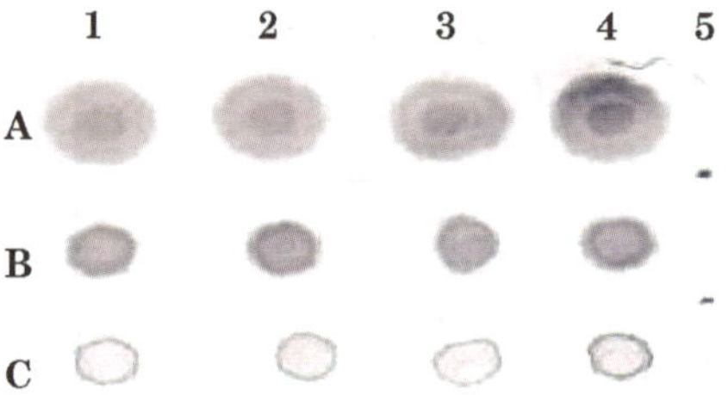

Fig. 58. Detection of ICRSV in Dot immunobinding assay (DIBA). Sample 1-4 are different isolates of ICRSV, sample 5 is healthy. A=1:1 dilution, B=1:10 dilution, C=1:20 dilution

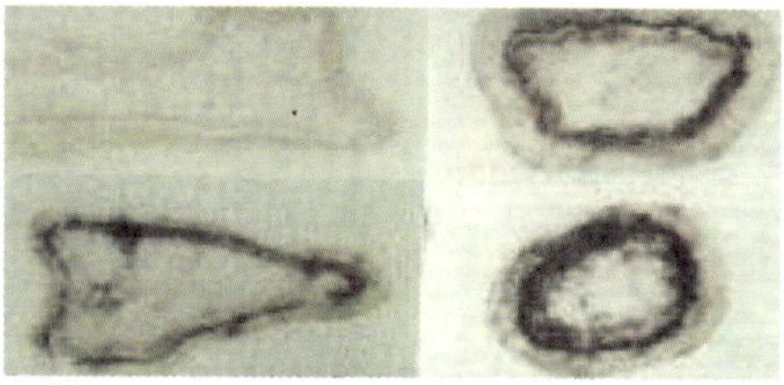

Fig. 59. Direct tissue blot immunoassay for citrus tristeza virus (CTV). Freshly cut surface of stem is blotted on NCM and exposed to CTV-specific antibodies which are directly or indirectly labeled with alkaline phosphatase and detected by exposer to a NBT-BCIP substrate. Note the reaction of positive samples in phloem tissue. *Source*: Garnsey *et al*. Proc. 12th IOCV.

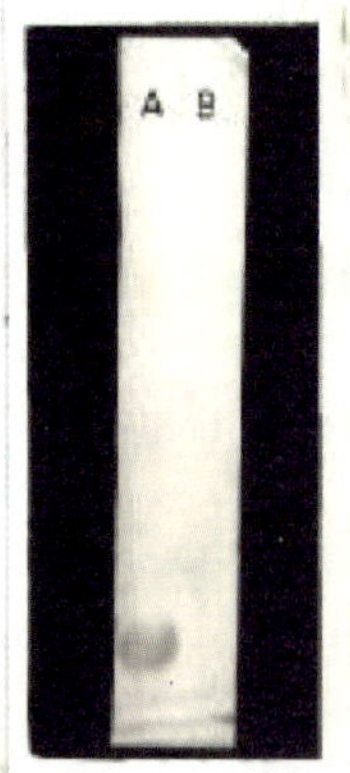

Fig. 60. Western blotting of ICRSV. A is a positive sample and B is control.

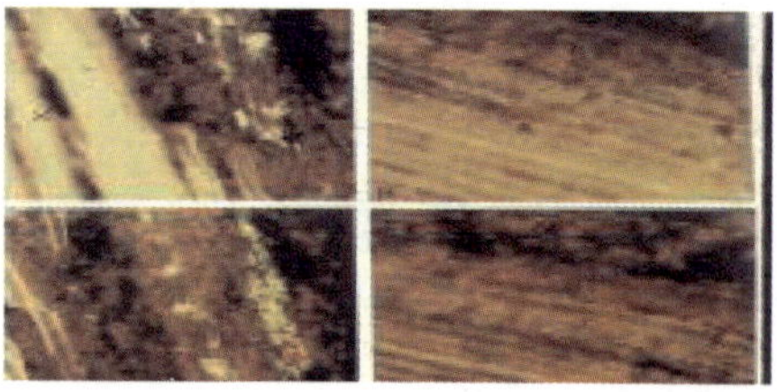

Fig. 61. Immunofluorescence test for detection of phloem limited viruses, phytoplasmas and citrus greening bacterium in ultrathin section. Note immunofluorescence in phloem tissues (left). Right are the healthy sections.

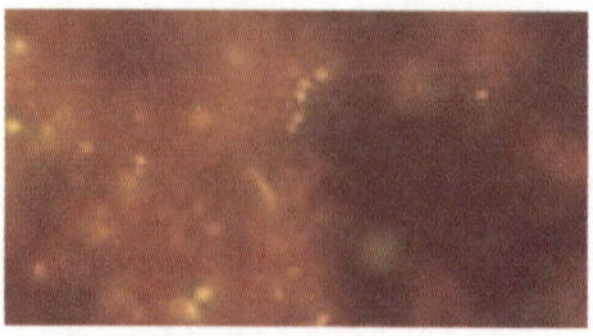

Fig. 62. Immunofluorescence test in extract of citrus greening affected leaf midribs. Note elongated and rounded forms of bacterium.

PLATE-XV

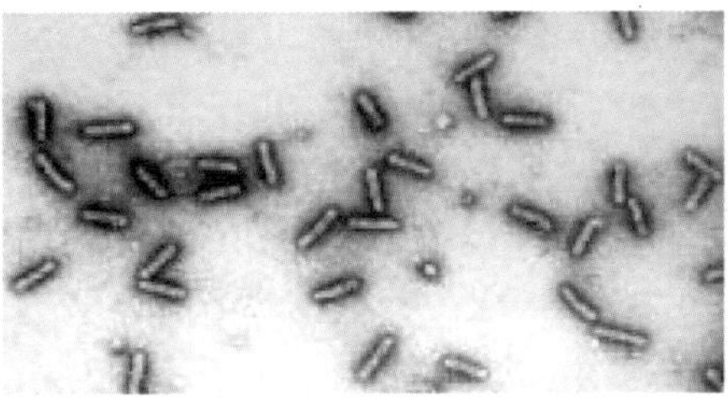

Fig. 63. Trapping of bacilliform particles of citrus yellow mosaic virus using virus-specific antibodies

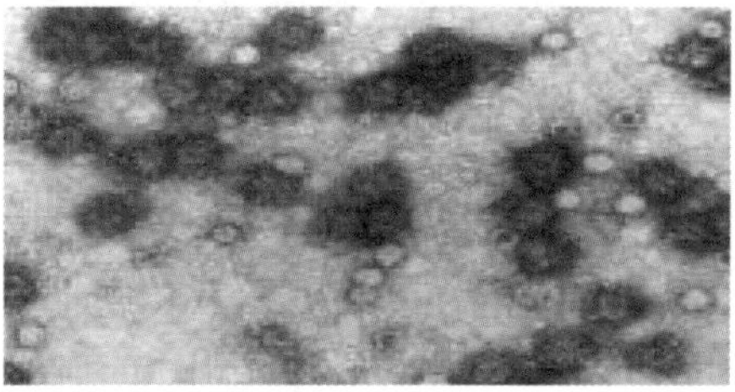

Fig. 64a. Decoration test of a sample having two viruses with spherical morphology. The antibodies specific to one virus decorated only to that virus and other virus remained undecorated as has been shown in the photograph.

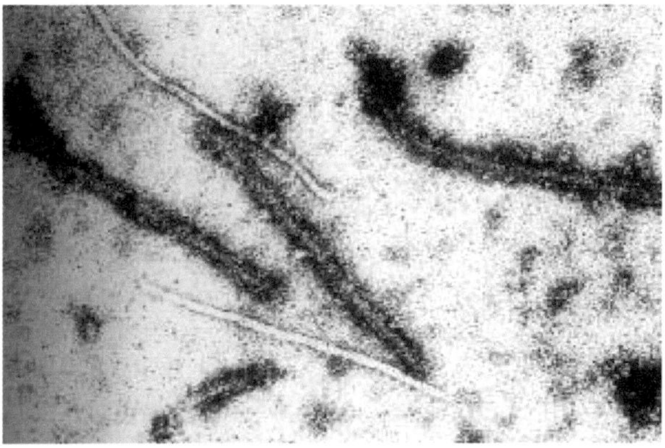

Fig. 64b. Similar result of decoration has been shown with filamentous viruses. Note decorated and undecorated virus particles.

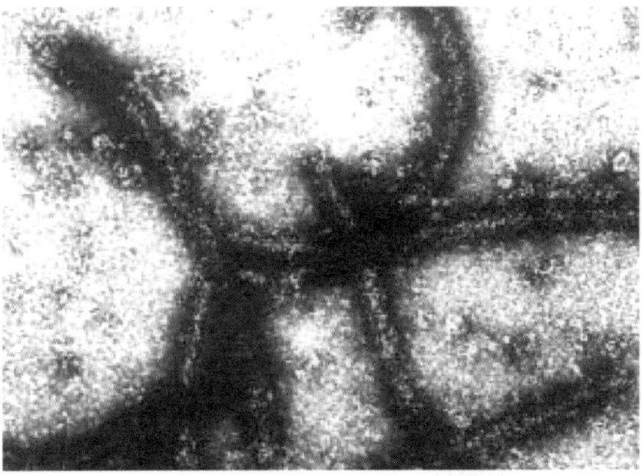

Fig. 65. The photograph shows the decoration of all the particles with homologous antiserum suggesting that the preparation had only one virus.

PLATE-XVI

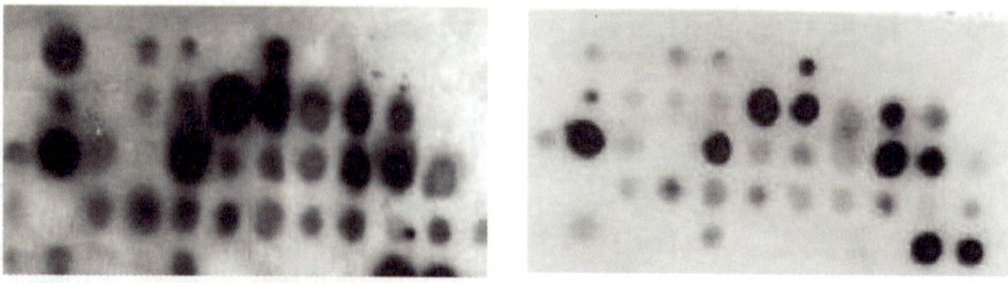

Fig. 66. Detection of Citrus greening bacterium by DNA-DNA hybridization using 2.6 kb probe. Left is at high exposer and right is at low exposer.

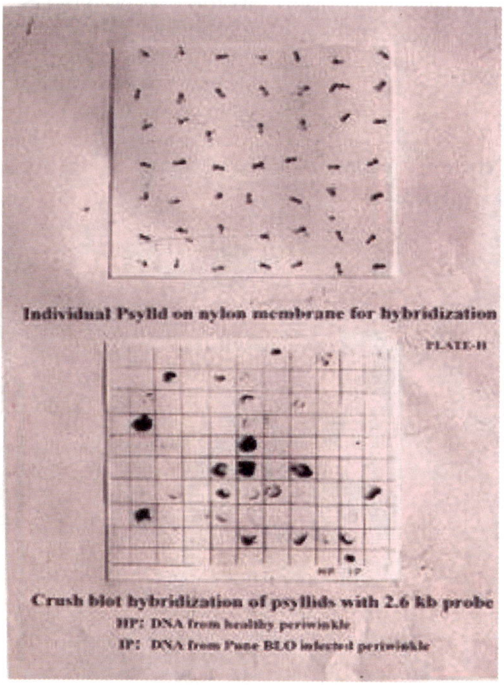

Fig. 67. Detection of citrus greening bacterium in citrus psylla, *Diaphorina citri* by DNA-DNA hybridization.

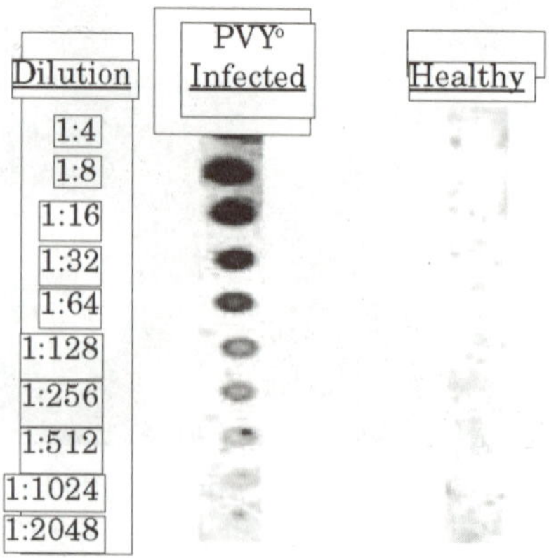

Fig. 68. Detection of PVY with DNA-RNA hybridization using Non-radioactive probe. Right is a control. See the clear detection upto a dilution of 1:256 of infected plant extract.

PLATE-XVII

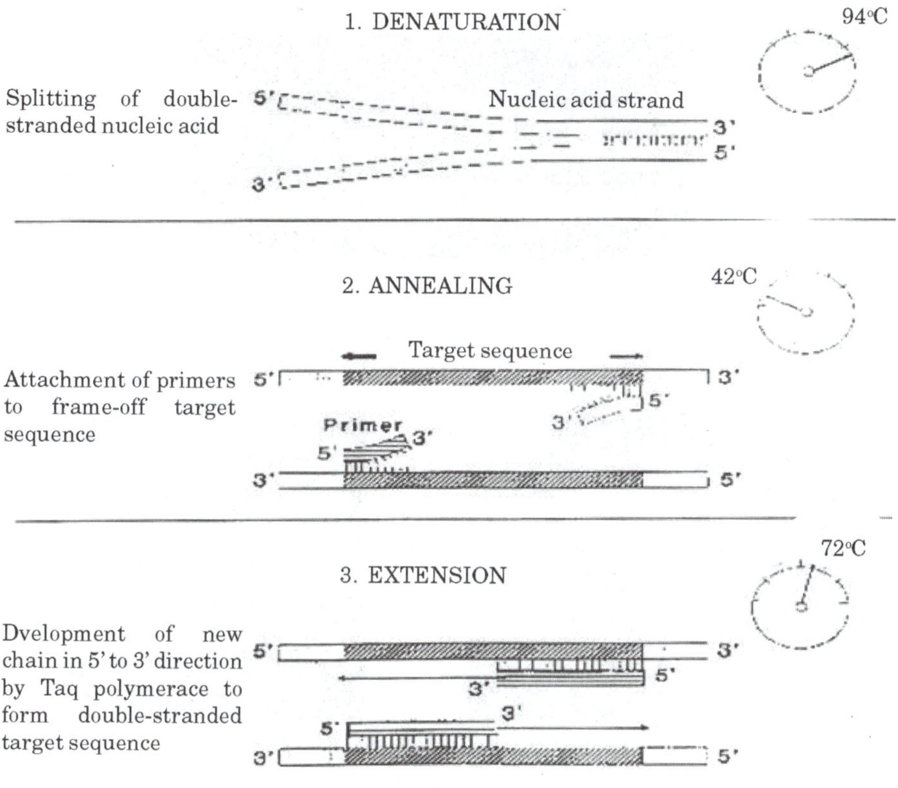

Fig. 69. Schematic presentation of PCR steps.

Fig. 70. RT-PCR of potato virus Y. Photograph is showing band in infected samples. Lower is a Healthy and PVY infected plant (Source: R.P. Singh).

PLATE-XVIII

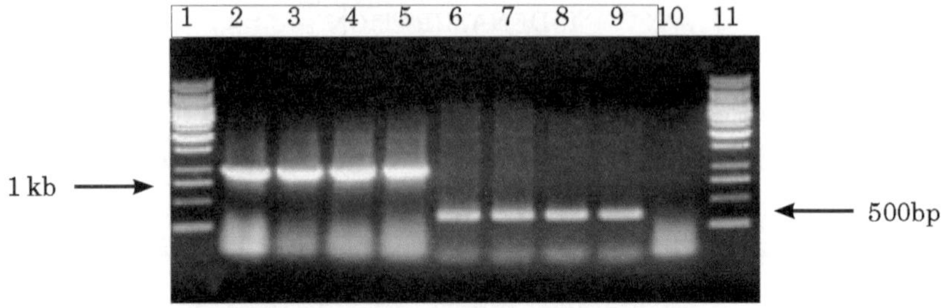

Fig. 71. PCR amplification of ICRSV CP gene with two sets of primers.
Lane 1 and 11 : 1k DNA ladder
Lane 2 and 6 : ICRSV Ab isolate
Lane 3 and 7: ICRSV Dl isolate
Lane 4 and 8: ICRSV Ah isolate
Lane 5 and 9 : ICRSV PU isolate
Lane 10: healthy control

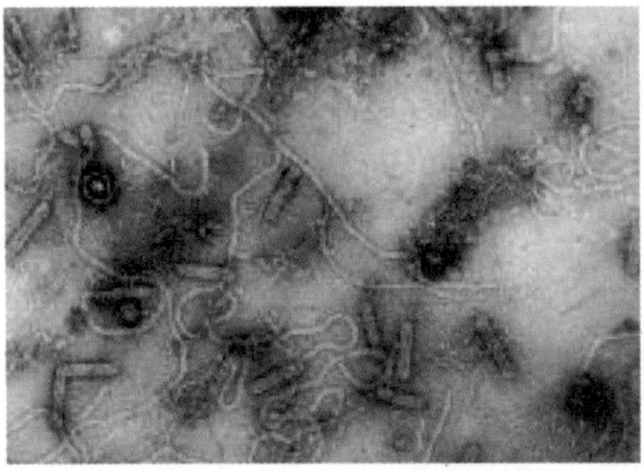

Fig. 72. Mixed infection of citrus viruses, citrus yellow mosaic virus (bacilliform particles), Indian citrus ringspot virus (filamentous particles) and highly flexuous particles of CTV.

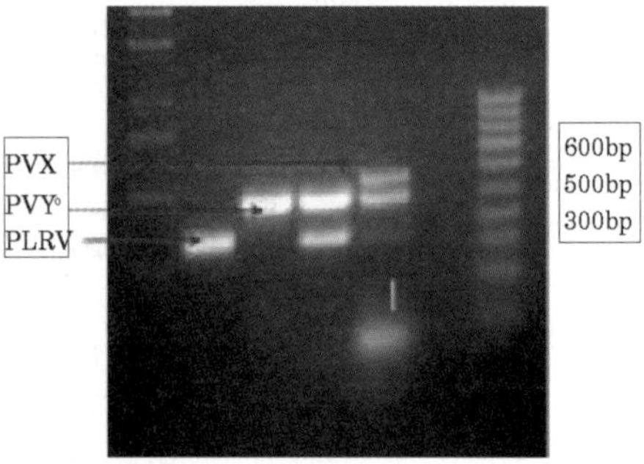

Fig. 73. Detection of potato viruses, Potato virus X (PVX), Potato virus Y (PVY) and Potato leafroll virus (PLRV).

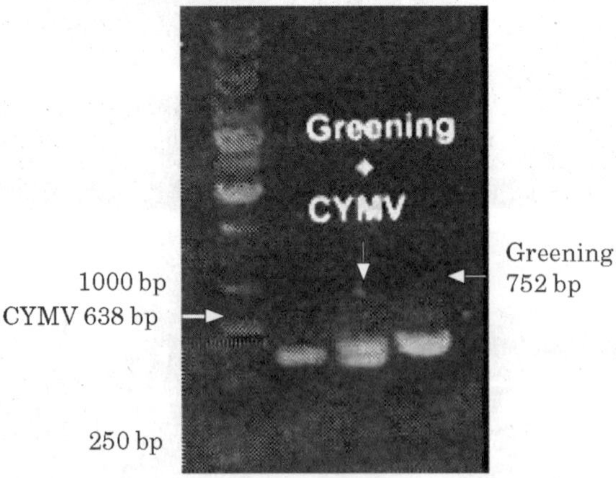

Fig. 74. Simultaneous detection of citrus yellow mosaic badnavirus and greening bacterium.

PLATE-XIX

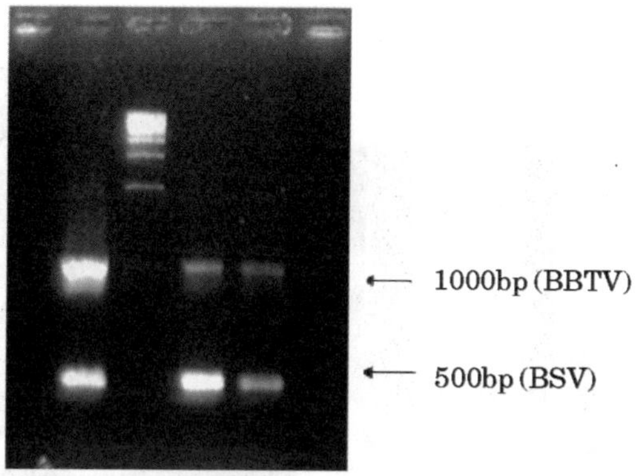

Fig. 75. Simultaneous detection of banana Bunchy top (BBTV) and streak virus (BSV)

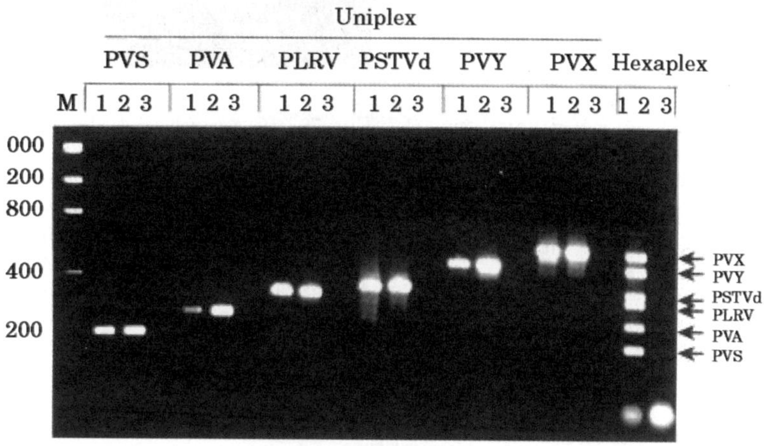

Fig. 76. Multiplex RT-PCR of 6 potato viruses. Source: R.P. Singh

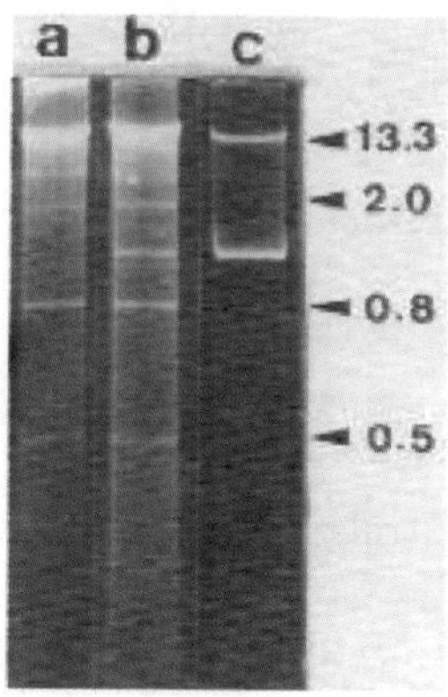

Fig. 77. dsRNA profile obtatined from pineapple Sweetorange plants inoculated with sub-isolates T-305 (a), on etrog citron plant inoculated with CTV isolate T-317 (b) or T-318 to Mexican lime (c). Sub-isolate T-305 and T-305a to Mexican lime. *Source*: Moreno *et al.*12th Proc. IOCV.

PLATE-XX

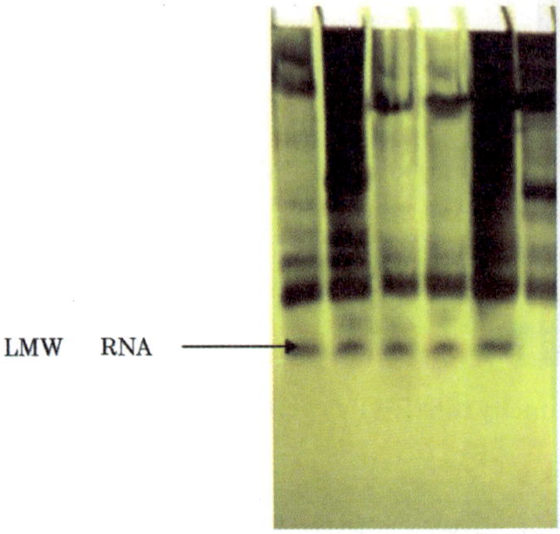

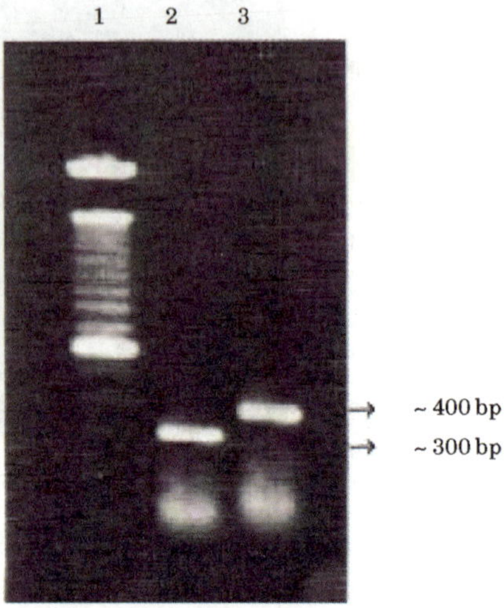

Fig. 78. Return – PAGE analysis of low molecular weight (LMW) RNA niprations showing bands at the extreme ends.Extreme right is a healthy control.

Fig. 79. 1% Agarose gel electrophoresis of RT-PCR amplified product from citrus yellow corky vein infected Kagzi lime plant. Lane 1: 100 bp DNA ladder Lane 2: Amplification with HSVd specific primer ~ 300 bp size of amplified product. Lane 3: Amplification with CEVd specific primer 400 bp size of amplified product.

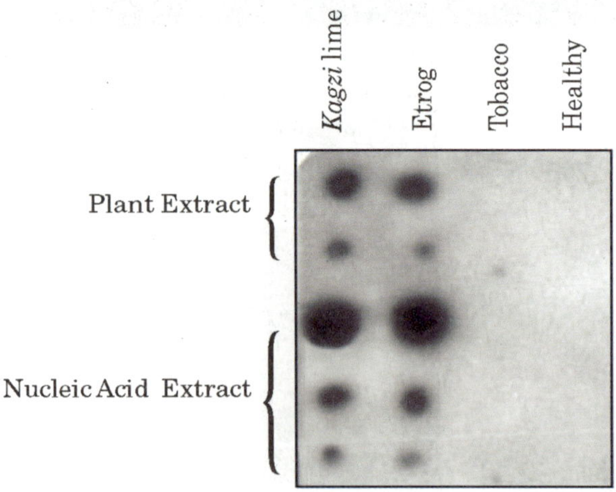

Fig. 80. Nucleic acid spot hybridisation with a cDNA probe of citrus exocortis viroid.

PLATE-XXI

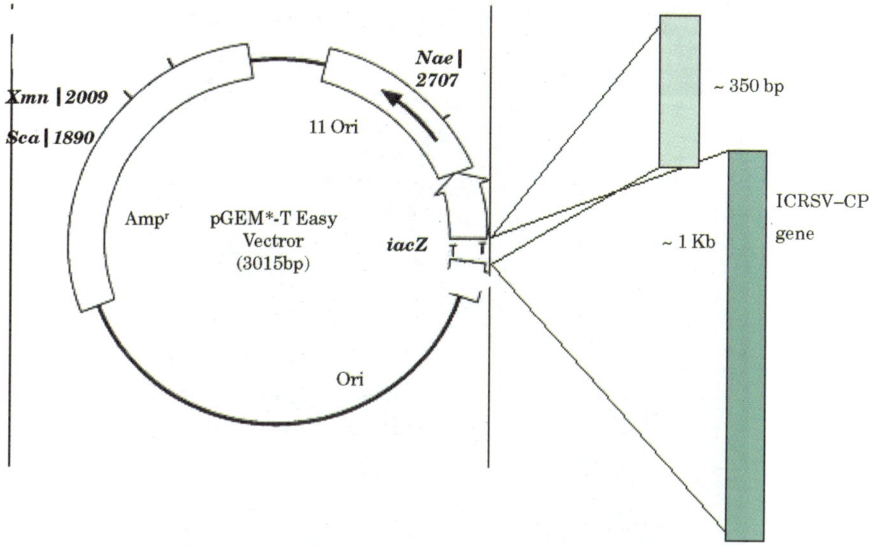

Fig. 81. Schematic of cloning strategy to amplify the two overlap fragments of coat protein gene of ICRSV on pGEM-T easy vector.

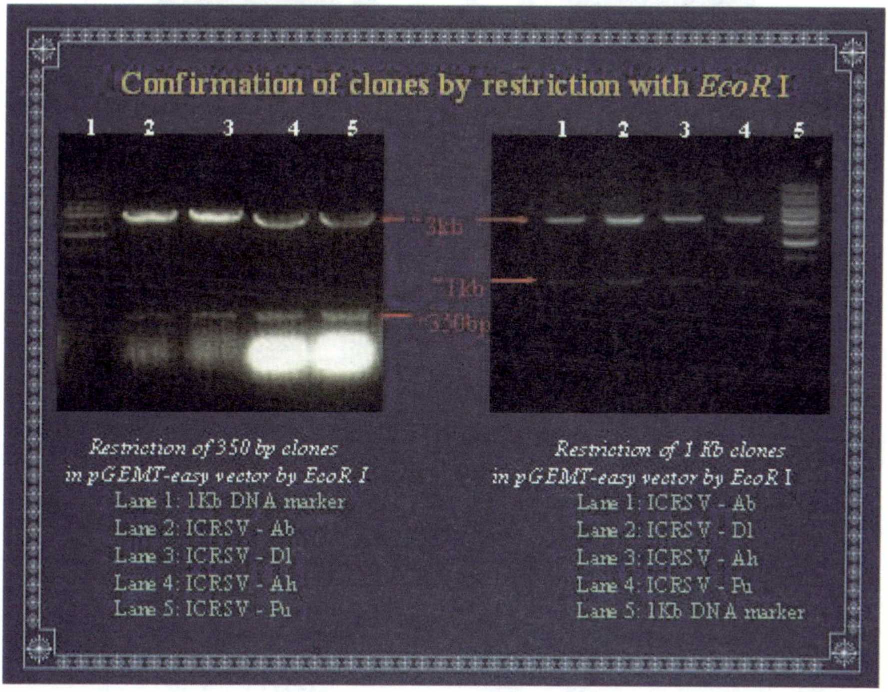

Fig. 82. (Left). Restriction of 350bp clone in vector by EcoRI.

Fig. 83. (Right). Restriction of 1kb clone in vector by EcoRI.

PLATE-XXII

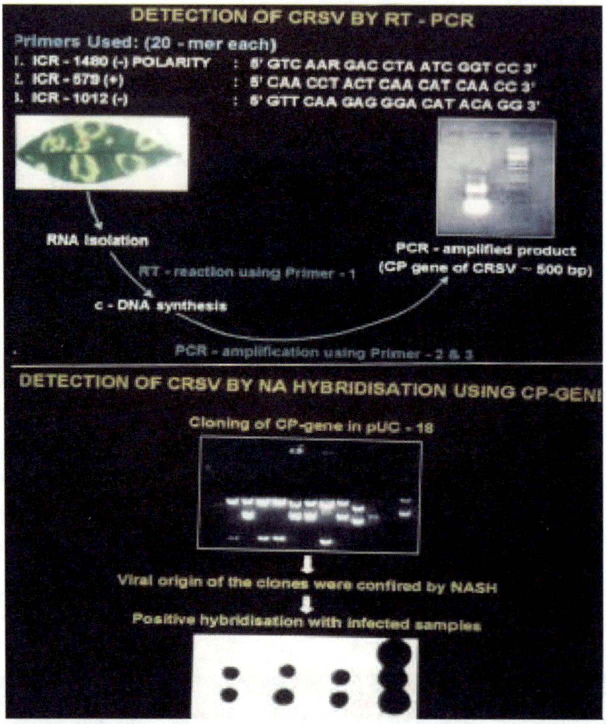

Fig. 84. The cloned fragment may be released by enzyme and may be used as probe for virus diagnosis in nucleic acid hybridization test.

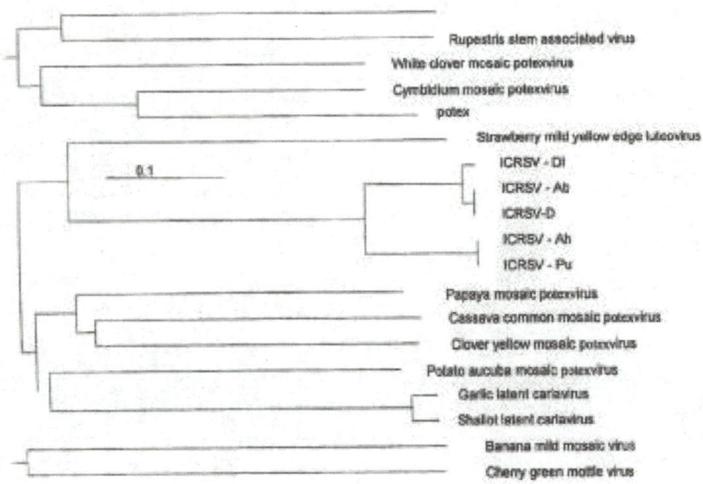

Fig. 85. Phylogenetic tree showing the relationship of ICRSV with other viruses.

PLATE-XXIII

Fig. 86. Banana plants being raised in tissue culture

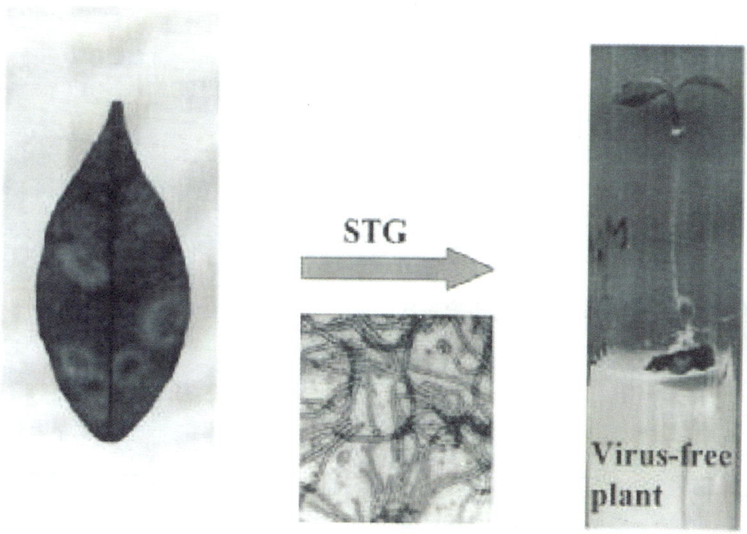

Fig. 87. A Kinnow mandarin plant raised by STG to clean it from ICRSV

PLATE-XXIV

Fig. 88. Ovule culture of kinnow mandarin

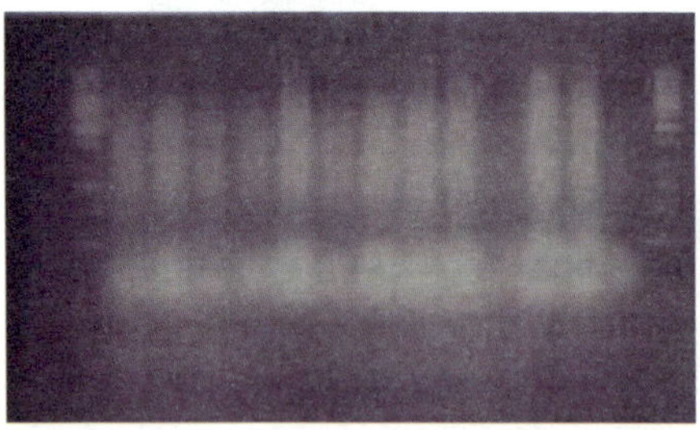

Fig. 89. Confirmation of gene construst in Agrobacterium by colony PCR from clones with pBI 121 using CP-FW/CP-R-Eco primers, most of the colonies showed smear PCR products

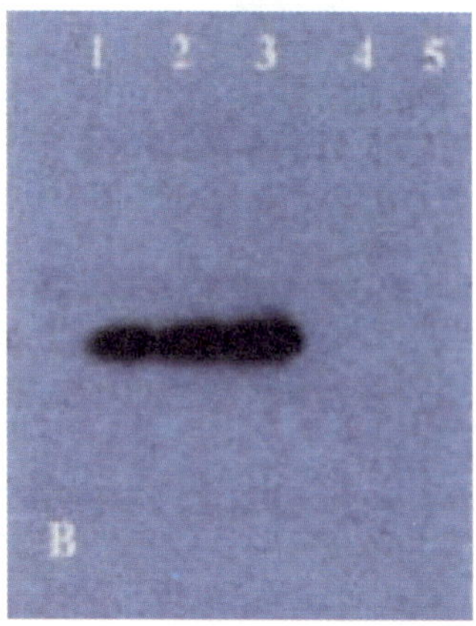

Fig. 90. Determination of transgene integration by southern hybridization using probe made from PCR product of ICRSV

CHAPTER 10

VIROIDS AND THEIR MOLECULAR DIAGNOSIS

There were many infectious virus-like diseases of plants where no virus could be isolated. It was in 1967 when Doi and his group from Japan reported that the mulberry dwarf disease was caused by mycoplasma-like organism (MLOs). Since then many 'yellows' diseases from where no virus could be isolated were turned out to be caused by MLOs (now known as phytoplasmas). However, still there were some diseases like potato spindle tuber from where neither a virus nor a phytoplasma could be established. It was in 1971 when Diener from USA isolated an infectious low-molecular weight RNA from potato spindle tuber and determined to be the causal agent for the disease and coined the term 'Viroid' to describe the infectious RNA.

About 30 viroids have been isolated from agricultural, horticultural and ornamental plants and fruit trees, including both monocotyledonous and dicotyledonous plants. Viroids have been isolated from plants with visible symptoms, as well as from those without any symptoms. Viroids can be defined as infectious, single stranded, covalently closed circular RNA of less than 400 nucleotides. Some of the important viroid diseases are, citrus exocortis, hop stunt, potato spindle tuber, peach latent mosaic viroid, chrysanthemum chlorotic mottle, avocado sun blotch, and coconut cadang cadang etc.

Since viroids are devoid of proteins, only nucleic acid based techniques have been developed for their detection. Currently, three types of detection techniques available based on the principle of gel electrophoresis, nucleic acid hybridization and genome amplification (PCR).

Isolation of total nucleic acid

1. Take 5 g leaf tissue from infected plant.
2. Wash, dry and cut into small pieces.
3. Homogenize in chilled pestle and mortar using 2 volume buffer (0.1 M Tris, 0.01 M EDTA), 1 M NaCl, 1% SDS containing 5 mM DTT and DIECA).
4. Add 3 volume chloroform: isoamyl alcohol (24:1), 3 Volume of Tris saturated phenol, pH 7.0.
5. Leave at room temperature for 10-15 min, shaking in between.
6. Centrifuge at 8500 rpm at 4°C for 15-20 min.
7. To supernatant add 0.5 volume ammonium acetate(7.5M, pH 7.0), 2.5 volume absolute alcohol.
8. Keep at -20°C overnight.
9. Centrifuge at 8500 rpm at 4°C for 20 min. The pellet is washed with 70% ethanol, dried and resuspended in 2 volume of extraction buffer. The above steps are repeated.
10. Dissolve pellet in sterile distilled water (100 µl/g tissue). Ready to test

Isolation of low molecular weight RNA (LMW-RNA)

Isolation of LMW-RNA from nucleic acid was done by column chromatography as described below.

1. Take CF-11 cellulose (Whatman) in proportion of 1 g/5 g fresh weight tissue.
2. Stirred well with 100 ml of sterile distilled water and kept undisturbed for 5 min.
3. Water is drained off and these steps are repeated twice.
4. The cellulose slurry in STE : ethyl alcohol is applied onto plastic columns with mira cloth filter at the end of the column.
5. Each column is then equilibrated with 10-12 ml STE: ethyl alcohol.
6. Samples were dissolved in STE : ethyle alcohol (1 ml/g fresh weight of tissue).
7. Load on top of the column gently and after it is absorbed, the column is washed with 10-12 ml of STE-ethyl alcohol until a clear elute is obtained.

8. Finally, nucleic acid is eluted by adding 2-3 ml of STE buffer and then elute as before.

Gel electrophoresis

Polyacrylamide vertical gels were used for electrophoretic separation of LMW-RNA under non-denaturing and denaturing conditions of run.

Non-denaturing conditions

1. LMW-RNA samples are electrophoressed in 5% polyacrylamide native gel (acrylamide : bisacrylamide 29:1) in vertical slabs in TBE buffer.
2. Carry out electrophoresis at room temperature with 1 x TBE as running buffer using a power source set at 46 MA (constant current) for 2.5 h.
3. Load 16 µl test nucleic acid in slots mixed with 4 µl of bromophenol blue-xylene cyanol (tracking dye) prepared in 40% glycerol-sucrose.
4. The LMW-RNA band can be seen in the gel.

Denaturing condition or return polyacrylamide gel electrophoresis (R-PAGE)

R-PAGE is bidirectional gel electrophoresis under denaturing conditions. The first run is done at 46 mA for 2-2.5 h using TE (Ix) buffer as running buffer. The same gel is rerun by replacing the TBE (Ix) with TBE (0.8X) at 70^{0}C in reverse polarity for 1.5-2 h. Electrophoresis apparatus is connected to water circulator bath set at 70°C and maintained so doing the entire run. The low ionic strength of the buffer and high temperature created denaturing conditions. The proptocol in brief is given below.

1. Approximately 20 µg of nucleic acid sample is electrophoresed through a 7.5% polyacrylamide slab gel (29:1 acrylamide: bisacrylamide) (15 x 16 x 0.12 cm) in 1 x TBE buffer (89 mM-Tris, 89 mM-boric acid, 2 mM EDTA) at 46 mA and 20^{0}C until the xylene cyanol is at the bottom of the gel.
2. The polarity of electrodes is reversed.
3. The buffer in the chambers is replaced with a 1:8 dilution of the 1 x TBE buffer at 70°C.
4. Electrophoresis is stopped after 1.5 h.
5. The gel is stained with silver nitrate to resolve the presence of nucleic acid. For silver staining, gels are immersed in fixative-1 immediately after the run was over and left at room temperature for 1 h or overnight. It is then transferred to fixative two for 30 min to 1 h at room temperature in a reciprocal shaker. The gel is stained with silver nitrate (0.2%) for 1h, developed in solution-1 and solution 2 and finally preserved in 5% acetic acid.

6. The gels can also be stained with 10 mg/ml ethidium bromide for 15 min. It is examined under photodyne transilluminator directly under UV light.
7. Use positive control, if available and also negative control.
8. A low molecular weight RNA band is observed (Fig. 78).

Detection of viroids by RT-PCR

The RNA isolated from infected leaves is used as template for cDNA synthesis using commercial kit (Quagen omniscript RT kit) or as described in earlier chapters. A set of primers is selected according to the specificity of the viroid, in question, and following protocol may be followed.

1. Template RNA is incubated at 70°C for 3-5 min.
2. Chilled on ice for 3-5 min.
3. Add 0.1 µg/ml reverse primer to the tube with other reaction components as described in PCR chapter.
4. The tube is chilled on ice and give a quick spin to bring down the contents.
5. For the first strand cDNA synthesis, the reaction is incubated at 37°C for 1 h.

The temperature profile and cycles used for PCR amplification are: 1 cycle for denaturation of DNA at 94°C for 2 min, 40 cycles which have three segments: denaturation at 94°C for 30 sec, annealing at 45°C for 30 sec, chain extension at 72°C for 30 sec. After PCR, 5 µl of amplified product was subjected to electrophoresis in 1% agarose gel to resolve the desired product. A band of amplified product can be observed with infected samples (Fig. 79).

Detection of viroids by nucleic acid hybridization

Like viruses, viroids can also be detected by nucleic acid hybridization test normally called as nucleic acid spot hybridization (NASH).

Preparation of blots for nucleic acid hybridization

1. 1 g leaf tissue from infected plant is ground in liquid nitrogen.
2. Plant sap and nucleic acid are extracted as described earlier.
3. Total nucleic acid isolated from healthy plant and PCR amplified product will serve as negative and positive control.
4. Nitrocellulose membrane (NCM) is cut according to requirement and 1cm squires are made over it by a soft lead pencil.
5. NCM is first soaked in distilled water and then in 10 x SSC for 5 min.
6. Excessive SSC is drained out and NCM is put on a filter paper for dotting.

7. The prepared samples are heat denatured by keeping them in boiling water in a water bath for 5 min. and then dotted on NCM
8. After dotting, the membrane is dried and baked at 80°C for 2 h.

Probe production

1. PCR amplified viroid components are electrophoresed in low melting point (LMP) agarose gel (1%) at 40V in 1 x TAE buffer.
2. The LMP agarose block containing the DNA fragment is cut from the gel.
3. Put it into a water bath with boiling water for 5 min.
4. 20 µl of gel mix is taken in a sterile eppendorf tube.
5. Add 1 µl random hexanucleotide primer and boil for 1 min.
6. Reaction mixture for probe can be prepared by mixing the reagents using random primer labeling kit (Bangalore Genei Pvt. Ltd.) and radiolabeling is done with a^{-32}P dCTP by random priming method.
7. The reaction mixture is incubated at 37°C for 2 h.
8. Before hybridization, double stranded ^{32}P-labeled probe is denatured in boiling water in a waterbath for 5 min.

Prehybridization and hybridization

1. Prehybridization solution (5X Denhardt's reagent, 6 x SSC, 0.3% SDS, 100 µg/ml^{-1} yeast RNA) is added to NCM at the rate of 0.2 ml/sq cm.
2. The cylinders are then incubated at 65°C for 4 h in hybridization oven with gentle rotation.
3. The denatured probe is added to prehybridization solution at 0.5 x 10^6 cpm/ml concentration.
4. Allow hybridization for 18 h at 65°C in hybridization oven with gentle rotation.
5. Hybridization solution is discarded and membrane is washed three times with 2 x SSC and 0.1% SDS at 65°C, each time for 15 min.
6. Washed membrane is dried on a paper towel, then place it within folds of cling film, place in a lead cassette and exposed to X-ray film (Kodak) for 18 h at -80°C.
7. Develop autoradiograph according to manufacturer's instruction.
8. The samples with positive hybridization can be seen on the membrane (Fig. 80).

Requirements

CF-11 Column Chromatography

STE buffer

0.1 M sodium chloride, 0.001 M EDTA, 0.05 M Tris HCl, adjust the pH to 7.5 with HCl and make up the volume to 10 ml.

Polyacrylamide gel electrophoresis (PAGE)

Acrylamide solution

30.0 g Acrylamide, 0.8 g N,N'-methylene bisacrylamide, 100 ml distilled water, filter through Whatman filter paper No.1.

TBE buffer

89 mM Tris, 89 mM boric acid, 2.5 mM EDTA, pH 8.3 (adjust with HCl), make up the volume.

TEMED (N,N,N',N' tetramethylethylenediamine)

Electrophoresis grade, ammonium persulfate APS) add in freshly prepared 10% (W/V) stock solution.

Gel solution (5%)

Distilled water	28.95 ml
Acrylamide-bisacrylamide	6.66 ml
10 x TBE buffer	4.00 ml
TEMED	40.00 µl
APS (10%)	350 µl
Total	40 ml

Loading dye

Bromophenol blue	0.5%
Xylene cyanol	0.5%
Glycerol	40%

Silver Nitrate staining

Fixative 1

5% glacial acetic acid	23.50 ml
25% ethyle alcohol	125 ml
Double distilled water(DDW)	88.50 ml

Fixative II

5% Glacial acetic acid	2.35 ml
25% Ethyle alcohol	25.00 ml
DDW	189.65 ml

STAIN

Developer

Silver nitrate 0.41 g dissolved in 200 ml DDW.

Solution 1

Potassium hydroxide	8.4 g
37% formaldehyde	1.51 ml
DDW	198.49 ml

Developer

Solution 2

1.48 g of sodium carbonate is dissolved in 200 ml of DDW.

Preservative

5% glacial acetic acid

Staining procedure

1. Immerse gel in fixative 1 and leave at room temperature for 1 h in a reciprocal shaker or overnight.
2. Transfer the gel to fixative II for 30 min-1 h at room temperature.
3. Stain the gel with silver nitrate (0.2%) for 1 h.
4. Wash 2-4 times with sterile DDW.
5. Transfer to Developer solution 1 for 2-3 min until golden brown color begins to appear.
6. Transfer to developer solution 2 for 1-2 min until resolution of bands is clear.
7. Preserve the gel in 5% acetic acid.

CHAPTER 11

CLONING AND SEQUENCING

Cloning of viral coat protein gene

The PCR amplified DNA is excised and eluted from the gel using QIAquick gel extraction kit (Qiagen Gmbh, Hilden, Germany). The purified PCR product is cloned and ligated into $p^{GEM\text{-}T}$ easy vector (3015 bp) from Promega (Fig. 81). The viral DNA is restricted with restriction enzyme EcoR 1. The presence of the insert in recombinant clones is confirmed through restriction analysis, miniprep DNA and colony PCR. Selected clones are sequenced at the automated DNA sequencing facility at any place for example in the Department of Biochemistry, Delhi University, south Campus, New Delhi, India or at Banglore Genei DNA sequencing facility using the ABI Prism BigDye terminator cycle sequencing protocol (Applied Biosystems, Perkin-Elmer Corp). Sequence analysis may be done using the computer program CLUSTAL X and GeneDoc version 2.6.002 and the sequence identity matrix for pair-wise combination of aligned sequences is calculated with Bio Edit Sequence Alignment Editor. The phylogenic relationship of the CPG sequences with exotic and indigenous isolates is generated using the programme Tree View (Win 32) version 1.6.6. The complete sequences of CP gene may be deposited to GenBank database and accession number obtained. The detail of various techniques involved in cloning and sequencing are given under following heads.

Ligation

The purified DNA getting after gel extraction is used for ligation as per the following composition:

PCR amplified product (50 ng/µl)	3 µl
$p^{GEM\text{-}T}$ easy vector (50 ng/µl)	1 µl
10X ligase buffer	1 µl
10 mM ATP	1 µl
T4 DNA ligase (0.2U/µl)	1 µl
Sterile distilled water	3 µl
	10 µl

In addition to two sets of standard reaction mixture, positive and negative control mixes are prepared by adding control insert and no DNA in ligation mixture respectively. The ligation mixtures are incubated at 4°C for 18 h. Both the ligation mixtures were then ready for transformation.

Preparation of competent cells

The competent cells are prepared as mentioned below:

1. 50 ml Luria broth (LB) is inoculated with 500 µl overnight grown culture of DH5a strain of *Escherichia coli* (*E. coli*) (Stretegene) and incubated at 37°C for 90 min with constant shaking at 200 rpm in a shaker incubator till the bacterial growth as measured by optical density reaches 0.3 OD at 600 nm.
2. The culture is then aseptically transferred to 40 ml sterile screw capped tubes and kept on ice for 10 min.
3. The cells are centrifuged at 500 rpm for 10 min at 4°C to palletize the cells.
4. The cells are resuspended gently in 10 ml ice cold 0.1 M $MgCl_2$ solution and kept on ice for 1 h.
5. The pellet is resuspended in 10 ml ice cold 0.1 M $CaCl_2$ solution and kept on ice for 1 h.
6. The cells are recovered by centrifuging at 5000 rpm for 10 min at 4°C and the pellets are resuspended in 1 ml of chilled 0.1 M $CaCl_2$ and use for transformation after keeping on ice for 1 h.

Transformation of competent cells

1. 200 µl competent cells are added to 10 µl of the ligation mixture in a sterile microfuge tube, mix gently and keep on ice for 1 h.

2. The bacterial cell DNA mixture is given a heat shock at 42°C for 2 min. One ml of LB medium is then added and the transformants are allowed to grow at 37°C for 1 h in shaker incubator at 200 rpm.
3. 200 µl of cell suspension is aseptically plated either as such or after concentration on Luria Agar (LA) plate containing 100 µg/ml of ampicillin, 200 µg/ml of X-gal, and 20 µl of 0.1 M IPDG for 100 ml LA.
4. The plates are incubated overnight at 37°C.

Selection of Transformants

The transformants are selected on the basis of blue/white colony color. The white colonies (transformants) are selected and subsequently plated on IPTG-X-gal-ampicillin plates. The plate having individual transformants in grid served as master plate.

Rapid screening for the recombinant clones having insert

White recombinant bacterial colonies may be formed either due to insertion of viral genome or due to even a small deletion in the Lac Z operon of vector itself. To know which of the white colonies have the insert, a rapid screening method is used as described below:

1. A small amount of overnight grown bacterial colonies are picked up from master grid individually with the help of sterile tooth pick and mixed with 50 µl of 10 mM EDTA (pH 8.0) in sterile microfuge tubes. One or two blue colonies of $p^{GEM\text{-}T}$ transferred bacterial cells are also taken for control.
2. 50 µl of fresh lysis solution (2N NaOH, 0.5% SDS, 20% sucrose) is added in each tube and vortexing is done for 30 sec which causes rapid disruption of bacterial cell wall.
3. The mixture is then incubated at 70°C for 5 min and is cooled down to room temperature.
4. 1.5 µl of 4 M KCl and 0.5 µl of 0.4% of bromophenol blue is added to each tube and vortex is done for 30 sec.
5. The mixture is then incubated for 5 min on ice and centrifuge is done at 10,000 rpm for 30 min at 4°C in bench top centrifuge.
6. Bacterial cell debris is removed and 30 µl of supernatant from each tube along with control are electrophoresed in 0.8% agarose gel.
7. The lanes which showed higher plasmid band than the control is considered as recombinant plasmid and the respective colonies are taken for isolation of recombinant plasmid DNA.

Isolation of recombinant plasmid DNA by miniprep method

1. Selected white colonies, which are presumed to have viral insert in rapid screening are individually inoculated in 2 ml of LB medium containing ampicillin (50 µg/ml) in sterile capped culture tubes.
2. Tubes are incubated overnight at 37°C at 200 rpm in a shaker incubator.
3. The overnight grown bacterial cells are then transferred to 1.5 ml sterile eppendorf tube and cells are harvested by centrifugation in a bench top centrifuge for 1 min. Care is taken to remove the medium adhering to the cell pellet.
4. The pellet is resuspended in 100 µl of solution I and mixed vigorously by vortexing.
5. Then 200 µl of freshly prepared lysis solution *i.e.* solution II is added and mixed gently.
6. 150 µl of ice cold solution III was then added and mixed gently with lysed cell suspension and the mixture is kept on ice for 15 min.
7. The chromosomal DNA and the bacterial cell debris are removed by centrifugation at 15000 rpm at 4°C in a bench top centrifuge.
8. The supernatant is again centrifuged for another 15 min at 15,000 rpm to pellet any unwanted bacterial debris.
9. The supernatant is taken and added phenol : chloroform : isoamyl alcohol in a ratio of 25:24:1. It is vortexed well and centrifuge in a bench top centrifuge at room temperature.
10. The clear aqueous phase is transferred to fresh eppendorf tube.
11. The DNA in aqueous phase is precipitated by adding 0.8 volume of isopropanol, keep on ice for 10 min.
12. Centrifuge the mixture at 15,000 rpm for 20 min at 4°C.
13. 200 µl of 70% ethanol is added to the pellet. The tube is rotated well so that the pellet from the wall gets suspended in 70% alcohol. DNA is then palletized by centrifugation at 15,000 rpm for 15 min.
14. The pellet is finally suspended in 30 µl sterile distilled water.

Large scale plasmid DNA preparation

Large scale plasmid DNA preparation of selected CPG clones can be obtained following QIAGEN midi plasmid purification protocol. The procedure as given in the manual should be followed.

Plasmid isolation buffers

Solution I (Resuspension buffer) 25 mM Tris HCl, pH 8.0

50 mM Glucose

10 mM EDTA

Solution II (lysis buffer) 0.2 N NaOH

1% SDS

Solution III (Neutralization buffer) 3 M Sodium acetate pH 4.8

Release of Insert with restriction Enzyme

Recombinant plasmids are subjected to digestion with restriction enzyme. Any enzyme is so chosen as the restriction site of this enzyme is present in both side of insertion site of vector but not present in the insert as known from the earlier reported sequence.

Restriction mixture is incubated at 37°C for overnight for complete digestion. Restriction is done to release the insert and also to know the insert size.

The reaction mixtures are prepared for each of the restriction digestion as follows:

Recombinant plasmid DNA	1.0 µl
10 X reaction buffer	2.5 µl
Restriction enzyme (10U/µl)	0.5 µl
Sterile distilled water	21.0 µl
Total	25 µl

After restriction digestion, the products are electrophoresed in 1% agarose gel. Fragment size was assessed in comparison with 1 kb DNA ladder as molecular weight marker on the same gel along with the DNA samples. The results of restriction of clones are given in Figs. 82 and 83.

Restriction analysis

For analysis of the restriction sites of the CPG, the sequences of the virus isolate are analysed using clone map software (version 2.11), the common restriction sites are chosen and may be shown on the map along with those common to all isolates of the virus. Restriction maps are confirmed by using common restriction enzymes, Pst 1, sac 1, and Cla 1 to digest the PCR products.

The clone after release by restriction enzyme may also be used as probe for dot blot hybridization (Fig. 84).

Nucleotide sequencing

Selected recombination clone with the insert and purified PCR product can be sequenced at places mentioned above. Sequences were aligned, translated and compared with the sequences of virus isolates published earlier using the MacVector program (BLAST) and Clustal W (version 1.82) multiple sequence alignment. Cluster dendrograms with pair-wise distance matrices are also generated.

The amino acid sequences are also used for blast separately in internet, the matching points with some other viruses in various groups can be observed. The amino acid sequences of such viruses are used for multiple sequence alignment in Bioedit. The position and percentage of matching can be determined. This multiple sequences alignment is also used to draw a phylogenetic tree using Treecon software. For example a phylogenetic tree is shown in Fig. 85.

CHAPTER 12

MICROARRAYS – AN INTRODUCTION

The proper and harmonious expression of a large number of genes is a critical component of normal growth and development and the maintenance of proper health. Disruption of changes in gene expression is responsible for many diseases. Microarrays allow scientists to analyze expression of many genes in a single experiment quickly and efficiently. They represent a major methodological advance and illustrate how the advent of new technologies provide powerful tools for researchers. Although microarrays are in its infant stage in plant sciences but scientists are using this technology to try to understand fundamental aspects of growth and development as well as to explore the underlying genetic causes of many human diseases.

DNA MICROARRAYS

Two recent complementary advances, one in knowledge and another in technology, are greatly facilitating the study of gene expression and the discovery of the roles played by specific genes in the development of diseases.

A microarray is a tool for analyzing gene expression that consists of a small membrane or glass slide containing samples of many genes arranged in a regular pattern.

Why are Microarrays Important?

Microarrays are a significant advance both because they may contain a very large number of genes and because of their small size. Microarrays may be used to assay gene expression within a single sample or to compare gene expression in two different cell types or tissue samples, such as in healthy and diseased tissue. This technology is still considered to be in its infancy; therefore, many initial studies using microarrays have represented simple surveys of gene expression profile in a variety of cell types. With new advances, researchers will be able to infer probable functions of new genes based on similarities in expression patterns with those of known genes. The use of microarrays may also speed the identification of genes involved in the development of various diseases by enabling scientists to examine a much large number of genes.

WHAT IS A DNA MICROARRAY?

DNA microarrays are small, solid supports on to which the sequences from thousands of different genes are immobilized or attached, at fixed locations. The supports themselves are usually glass microscope slides, the size of two side-by-side pinky fingers, but can also be silicon chips or nylon membranes. The DNA is printed, spotted, or actually synthesized directly on to the support. The spots can be DNA, cDNA, or oligonucleotides.

An oligonucleotide, or oligo as it is commonly called, is a short fragment of a single-stranded DNA that is typically 5-50 nucleotides long.

DESIGNING A MICROARRAY EXPERIMENT

How will you extract information about a disease condition from a dime-sized glass or silicon chip containing thousands of individual gene sequences? The whole process is based on hybridization probing, a technique that uses fluorescently labeled nucleic acid molecules as "mobile probes" to identify complementary molecules, sequences that are able to base-pair with one another. Each single stranded DNA fragment is made up of four nucleotides, adenine (A), thymine (T), guanine (G) and cytosine (C), that are linked end to end. Adenine is the complement of, or will always pair with thymine, and guanine is the complement of cytosine. Therefore, the complementary sequence to G-T-C-C-T-A will be C-A-G-G-A-T. When two complementary sequences find each other, such as the immobilized target DNA and the mobile probe DNA, cDNA or mRNA, they will lock together, or hybridize.

Now consider two cells: cell type 1, a healthy cell, and cell type 2, a diseased cell. Both contain an identical set of four genes, A,B,C, and D. Scientists are interested in determining the expression profile of these four genes in the two cell types. To do this, Scientists isolate mRNA from each cell type and use this mRNA as templates to generate cDNA with a "fluorescent tag" attached. Different tags (red or green) are used so that the samples can be differentiated in subsequent steps. The two labeled samples are then mixed and incubated with a microarray containing the immobilized genes A, B, C, and D. The labeled molecules bind to the sites on the array corresponding to the genes expressed in each cell. A DNA microarray experiment can be set up as below:

1. Prepare your DNA chip using your chosen target DNAs.
2. Generate a hybridization solution containing a mixture of fluorescent labeled cDNAs.
3. Incubate the hybridization mixture containing fluorescent labeled cDNAs with your DNA chip.
4. Detect bound cDNA using laser technology and store data in a computer.
5. Analyse data using computational methods.

After the hybridization step is complete, place the microarray in a "reader" or "scanner" that consists of some lasers, a special microscope, and a camera. The fluorescent tags are excited by the laser, and the microscope and camera work together to create a digital image of the array. These data are then stored in a computer, and a special programme is used either to calculate the red-to-green fluorescence ratio or to subtract out back ground data for each microarray spot by analyzing the digital image of the array. If calculating ratios, the programme then creates a table that contain the ratios of the intensity of red-to-green fluorescence for every spot on the array. For example, using the scenario outlined above, the computer may conclude that both cell types expresses gene A at the same level, that cell 1 expresses more of gene B, that cell expresses more of gene C, and that neither cell expresses gene D. Some microarray experiments can contain up to 30,000 target spots. Therefore, the data generated from a single array can mount up quickly.

THE COLORS OF MICROARRAY

GREEN: represents control DNA, where either DNA or cDNA derived from normal tissue is hybridized to the target DNA.

RED: represents sample DNA, where either DNA or cDNA is derived from diseased tissue hybridized to the target DNA.

YELLOW: represents a combination of control and sample DNA, where both hybridized equally to the target DNA.

BLACK: represents areas where neither the control nor sample DNA hybridized to the target DNA.

Each spot on an array is associated with a particular gene. Each colour in an array represents either healthy (control) or diseased (sample) tissue. Depending on the type of array used, the

location and intensity of the color will tell us whether the gene, or mutation, is present in either the control and/or sample DNA. It will also provide an estimate of the expression level of the gene(s) in the sample and control DNA.

TYPES OF MICROARRAYS

There are three basic types of samples that can be used to construct DNA microarrays, two are genomic and the third is "transcriptomic", that is, it measures mRNA levels. What makes them different from each other is the kind of immobilized DNA used to generate the array and, ultimately the kind of information that is derived from the chip. The target DNA used will also determine the type of control and sample DNA that is used in the hybridization solution.

I. Changes in gene expression level

During the level, or volume, at which a certain gene is expressed is called microarray expression analysis and the arrays used in this kind of analysis are called expression chips. The immobilized DNA is cDNA derived from the mRNA of known genes, and once again, at least in some experiments, the control and sample DNA hybridized to the chip is cDNA derived from the mRNA of normal and diseased tissue, respectively. If a gene is overexpressed in a certain disease state, then more sample cDNA, as compared to control cDNA, will hybridize to the spot representing that expressed gene. In turn, the spot will fluoresce red with greater intensity than it will fluoresce green. Once researchers have characterized the expression patterns of various genes involved in many diseases, cDNA derived from diseased tissue from any individual can be hybridized to determine whether the expression pattern of the gene from the individual matches the expression pattern of a known disease. Expression chips can also be used in disease diagnosis. Similar situation will also help to identify plant diseases.

II. Genomic gains and losses

Researchers use a technique called microarray comparative genomic hybridization (CGH) to look for genomic gains and losses or for a change in the number of copies of a particular gene involved in a disease state. In microarray CGH, large pieces of genomic DNA serve as the target DNA, and each spot of target DNA in the array has a known chromosomal location. The hybridization mixture will contain fluorescently labeled genomic DNA harvested from both normal (control) and diseased (sample) tissue.

Therefore, if the number of copies of a particular target gene has increased, a large amount of sample DNA will hybridize to those spots on the microarray that represent the gene involved in that disease, whereas comparatively small amount of control DNA will hybridize to those same spots. As a result, those spots containing the disease gene will fluoresce red with greater intensity than they will fluoresce green, indicating that the number of copies of the gene involved in the disease has gone up.

III. Mutation in DNA

When researchers use microarrays to detect mutation or polymorphism in a gene sequence, the target or immobilized DNA, is usually that of a single gene. In this case though, the target sequence place on any given spot within the array will differ from that of other spots in the same microarray, sometimes by only one or a few specific nucleotides. One type of sequence commonly used in this type of analysis is called a single nucleotide polymorphism (SNP), a small genetic change or variation that can occur within an individual's DNA sequence.

Once researchers have established that a SNP pattern is associated with a particular disease, they can use SNP microarray technology to test for that disease expression pattern.

Microarray data management

Use of microarray technology by many laboratories will make it difficult to manage the data as it will be tremendous. Suppose a chip contains 30,000 spots of target DNA, interpreting the data generated by that chip would need to know the biological identity of each target – what gene is where; the biological properties of the control and sample DNA; the experimental conditions and procedure used in setting up the experiment; and finally, the results.

To support the public use and dissemination of gene expression data, NCBI has launched the gene expression omnibus or GEO. GEO represents NCBI's effort to build an expression data repository and online resource for the storage and retrieval of gene expression data from any organism or artificial source.

Developing MAML: Reading off the same platform

Microarray markup language developed by the "MAML" working group of MGED, the Microarray Gene Expression Data base, is a first attempt to provide a standard platform for submitting and analyzing the enormous amounts of microarray expression data generated by different laboratories around the world. MAML proposes a framework for describing information about a DNA-array experiment and a data format for communicating this information, including its details.

- **Experimental design:** The set of the hybridization experiment as a whole.

- **Array design:** Each array used and each spot on the array.
- **Samples:** Samples used, the extract preparation, and labeling.
- **Hybridization:** Procedures and parameters.
- **Measurements:** Images, quantification, and specifications.
- **Control:** Types, values, and specification.

The benefits of GEO and MAML

- By storing vast amounts of data on gene expression profiles from multiple experiments using varied criteria and conditions, GEO will aid in the study of functional genomics – the development and application of global experimental approaches to assess gene function.
- GEO will facilitate the cross-validation of data obtained using different techniques and technologies and will help set benchmarks and standards for further gene expression studies.
- By making the information stored in GEO publicly available, the field of bioinformatics and functional genomics will be both promoted and advanced.
- That much experimental data should be freely accessible to all is consistent with NCBI's legislative mandate and mission : to develop new information technologies to aid in the understanding of fundamental molecular and genetic processes that control disease.

Microarrays provide hope for obtaining global view of biological processes and to open new avenues for molecular diagnosis and management of plant viruses in particular and plant pathogens in general.

CHAPTER 13

VIRUS ERADICATION FROM PLANTS

Scientists are trying to manage plant viruses for a long time. Some of the practical approaches developed so far are given below:

Meristem tip culture

The meristem tips of a targeted plant species are grown in a media based on Murashige and Skoog's medium (MS medium) as geven in Table 7.

Table 7. Culture medium for meristem tip culture

Mineral salts	mg/L	Organic ingradients	
Mn_4No_3	1650		
KNo_3	1900	Sucrose	30 g/L
$CaCl_2 2H_2o$	440	Indoleacetic acid	8 mg/L
$MgSO_4 7H_2O$	370	Kinetin	2.56 mg/L
KH_2PO_4	170		
Na_2-EDTA	37*	Pyridoxine HCl	0.5 mg/L
$FeSO_4.4H_20$	28*	Thiamine HCl	0.1 mg/L
H_3BO_3	6.2	Nicotinic acid	0.1 mg/L
$MnSO_4.4H_2O$	22.3	Myo-inositol	100 mg/L
$ZnSO_4.4H_2O$	8.6	Agar (Oxoid No.3)	9 g/L
KI	0.83		
$Na_2MoO_4.2H_2O$	0.25		
$CuSO_4.5H_2O$	0.025		
$CoCl_2.6H_2O$	0.025		

- 5 ml/L of a stock solution containing 5.57 g $FeSO_4.7H_2O$ and 7.45 g Na_2-EDTA per litre of distilled water

Students should try to follow the protocols if a plant species, in question, has already been grown *in vitro* by other workers. If no information is available, then try using the medium detailed in above table and carry out experiments using different concentrations of indoleacetic acid (IAA) and kinetin as shown in Table 8 below. These experiments may determine the optimum concentrations of auxin and cytokinin for the type of growth required.

Table 8. Examples of varying concentrations of auxin and cytokinin that may be used to induce different types of culture growth

Kinetin concentration#	Indoleacetic acid concentration#			
	0	**4 mg/l**	**8 mg/l**	**16 mg/l**
0	1*	2	3	4
1 mg/l	5	6	7	8
2.5 mg/l	9	10	11	12
10 mg/l	13	14	15	16

• Number of treatment (After Walkey,1985)

\# Other auxins and cytokinins may be tried instead of kinetin and indoleacetic acid

Generally, high cytokinin concentrations favour multiple shoot formation, and a cytokinin-free medium favours rooting. Other auxins and cytokinins may be substituted for IAA and kinetin. One litre of medium may be prepared as described below:

1. The mineral salts may be purchased as commercially pre-weighed prepared package and dissolve one sachet, 30 g sucrose and 100 mg myo-inocitol in 800 ml of distilled water.
2. Add to this solution.

Pyridoxin HCl : Dissolve 25 mg in a few ml of N/10 HCl and make up to 50 ml with distilled water (500 mg/L). Use 1 ml (0.05 mg/L) of this solution.

Thiamine HCl : Dissolve 25 mg in 50 ml of distilled water (500 mg/L). Use 0.2 ml (0.1 mg/L) of this solution.

Nicotinic acid : Dissolve 25 mg in a few ml of N/10 HCl and make up to 50 ml with distilled water (500 mg/L). Use 0.2 ml (0.1 mg/L) of this solution.

Indoleacetic acid: Dissolve 40 mg in a few ml of absolute alcohol and make up to 50 ml with distilled water (800 mg/L). Use 10 ml of this solution (8 mg/L).

Kinetin (6-furfurylaminopurine): Dissolve 25.6 mg in a few ml of N/10 HCl and make up to 50 ml with distilled water (512 mg/L). Use 5 ml (2.56 mg/L) of this solution.

3. Make up the total volume to 1000 ml with distilled water and adjust the pH 5.7 with N/1 NaOH.
4. Add 9 g of agar (oxoid No.3 or equivalent) and heat the solution until the agar is thoroughly dissolved.
5. Add 10 ml of molten solution to 7.0 x 2.5 cm flat bottomed, glass tubes. Seal each tube with an aluminum cap or foil.
6. Stack the tubes in a wire basket and cover the mouth of the basket with cotton wool and autoclavable paper.
7. Sterilize by autoclaving for 15-20 min at 15 lb pressure. The sterilized tubes may be stored at 3-4°C until used.

Heat treatment of the infected parent plant

Incubate the infected plant, growing in a pot in compost at 30-40°C in an illuminated, constant temperature growth cabinet (16 h photoperiod). The minimum treatment period is likely to be four weeks, but upto 12 weeks may be necessary. The plant should be treated for as long as possible at the highest temperature it can withstand. The meristem-tips must be removed immediately as soon as the high temperature treatment is completed.

Excision of meristem-tip

Sterilization of the explant

The shoot bearing the bud from which the meristem-tip is to be removed can be surface sterilized as follows:

1. Immerse shoot in 70% alcohol for 1-2 min and then immerse in 5% sodium or calcium hypochlorite solution containing 0.1% Tween-80 for 5 min.
2. Rinse several times in sterile distilled water for 3-5 min each rinse

Removal of the tip

Using a binocular stereomicroscope and a fine pair of foreceps to hold the shoot, remove the outer leaves of the bud with a sterilized needle or scalpel. Sterilize the needle by dipping in alcohol and flame between each cut, and remove the remaining primordial leaves until the youngest primordial leaf and meristem dome are exposed. Remove the meristem tip with a fragment of razor-blade mounted in a handle, and place the tip in a culture tube. The tip size should be 0.3-1 mm in diameter. After the tip is placed in the culture tube, the tube is plugged under aseptic and sterilized conditions. The culture should be grown at a constant temperature between 23 to 25°C and illuminated at 16 h per day.

Post-culture treatment

Transfer the rooted plantlets in to a 4 cm 'Jiffy' peat pot, and gradually 'harden off' the plant in a seed tray covered by a plastic cover in a glasshouse. Remove the plastic cover of seed tray after two weeks. If the plantlet is growing well, after a few days peat pot can be planted *in situ* into a pot containing compost. The plantlet should then be grown to maturity in sterilized soil in an insect-proof glasshouse and should be indexed for viruses.

Tissue culture raised plants are becoming popular among growers because of their better growth, true to type and better yield. Fig. 86 shows the banana plants being grown in tissue culture in a company.

Shoot tip grafting (STG) or micrografting

The technique has become very popular for multiplication of planting material in citrus. It is briefly given below:

Preparation of rootstock seedlings

Seeds of a suitable rootstock are peeled off, sterilized in a 0.1% mercuric chloride solution for 5 min and rinsed 3 times in sterile distilled water. The seeds are then grown in Murashige and Skoog medium, solidified with 1% agar (pH adjusted to 5.7) and the medium is autoclaved at 121°C for 15 min. The autoclaved medium is distributed in 25 ml aliquots in 25 x 150 mm culture tubes. One seed per tube is sown and culture tubes are maintained at 27°C. The seedling grown in dark become etiolated and may be better suited for micrografting.

Preparation of scion for shoot tips

Budwood containing lateral buds are collected from a tree with best horticultural qualities. All the leaves of the budwood are removed and budwood are sterilized in 0.1% mercuric chloride solution followed by washing 3 times by sterilized water. The budwood are then placed in culture tubes having medium as described earlier for growing. The cultures are maintained under 16 h daily exposures to 1000 lux light at a constant temperature of 37°C. The buds will start sprouting which are used as source of shoot tip.

Size of shoot tip

Normally size of shoot tip taken for STG is 2-3 leaf primordial but sometimes a size up to 6 leaf primordial is taken for higher success of STG.

The grafting procedure

1. A 2-week old seedling in a culture tube is decapitated, leaving 6 cm tall and 1.6-1.8 mm diameter at the point of decapitation. A cut is given to the stem at the top for insertion of shoot tip.

2. The root is shortened to 4-6 cm and cotyledons and auxillary buds are removed.
3. A shoot tip composed of the apical meristem and adjacent tissue is isolated from sprouting buds using a razor blade silver attached to a beaver surgical handle as a scalpel, and transferred to the cut of the decapitated stem.
4. The steps of shoot tip excision and transfer of rootstock are carried out aseptically under a laminar flow with the aid of a dissecting microscope

The shoot tip grafted plants are placed on MS medium with nutrients and vitamins, myoinocitol (100 mg/L), nicotinic acid (5 mg/L), Pyridoxin HCl (10 mg/L), thiamin HCl (10 mg/L), glysine (2 mg/L), sucrose (5%). 25 ml medium was placed with 1% agar.

The grafted plants are kept upright in dark for 5 days with a constant temperature of 27°C in a tissue culture facility. The adventitious shoots arising from rootstocks are removed aseptically with a forecep. After 5 days the light intensity is increased to 10,000 lux. A shoot tip grafted plant is shown in Fig. 87.

Transfer of STG plants to soil

Grafted plants showing 2-4 expended leaves in the shoot tip after 5-6 weeks of grafting are transplanted (with medium) directly to 10 cm pots having compost. The pots are covered with polythene bags to minimize moisture loss and shaded with brown paper for a week, and then the plants are allowed to grow in the glasshouse. The shoot arising from the grafted shoot tip may also be grafted directly to the root stocks in usual manner. After hardening, all STG plants must be tested for the presence of viruses preferable with DAC-ELISA with specific polyclonal antibodies

Note: All experiments of STG should be done under a laminar flow.

OVULE CULTURE

Both fertilized and unfertilized ovules can be cultured (Fig. 88). The flowers are collected of the targeted species and ovules are excised and placed on solidified medium in 12 cm Petri discs The medium to be used is either Murashige and Tucker (MT) alone or with benzyladenine (BA) and kinetin (KT) with a concentration of 10 mg/L containing 5% sucrose. The pH is adjusted to 5.7 and added 1% (W/V) agar and then autoclaved. The discs with ovules are kept in tissue culture room at 27°C with 16 h light.

The callus produced from the ovule culture is subcultured on MT medium with 1 mg/L of NAA or 2 mg/L of NAA + 2 mg/L of kinetin or with BA and gibberellic acid (GA) from 1-2 mg/L and 5% sucrose. All the cultures are maintained at 27°C with 16 h light.

Small *in vitro* cotyledonary embryos and plantlets can be used for grafting in fruit trees either by inverted T grafting (for embryo) or mini wedge grafting for plantlets. The grafted plants are enclosed in transparent plastic bags in order to conserve hugh hygrometry and maintained at 26°C with 16 h of dim light daily. After three weeks, the grafted plants can be shifted to glasshouse.

GENETIC ENGINEERING FOR MANAGEMENT OF PLANT VIRUS DISEASES

Genetic engineering brings new hope for the effective control of plant virus diseases. The concept of pathogen derived resistance (PDR) has stimulated research on obtaining virus resistance through genetic engineering. PDR is mediated either by the protein encoded by the transgene (protein-mediated) or by the transcript produced by the transgene (RNA-mediated). Many viral sequences have been engineered into the genome of a number of crops making them tolerant or resistant to specific viruses. Compared to conventional breeding for virus resistance, genetic engineering provides a quicker and more precise technology to obtain plants that are resistance to viruses.Recent researches indicate that PDR to viruses is mediated, in most cases, by an RNA-based post-transcriptional gene silencing mechanism.

The genes used for plant transformation are: DNA copies of satellite RNAs that reduce symptom expression, coat protein gene, non-structural gene sequences, defective interfereing molecules, ribozymes, anti-sense RNA, plantabodies, gene silencing and plant R-genes. In general, protein mediated resistance provides moderate protection against a broad range of related viruses while RNA-mediated resistance offers high level of protection only against closely related strains of a virus. To date, the most common type of PDR in advanced stages of development is coat-protein mediated resistance.

Pathogen derived resistance

Replicase-mediated and RNA-mediated resistance

In Replicase-mediated resistance, transgenic plants are highly resistant to infection, and in some cases, are essentially not infected by the virus from which the replicase gene is derived.

Movement protein-mediated resistance

Cell to cell and long distance spread of virus infection requires one or more viral protein(s). Therefore, based upon the role of movement proteins in virus disease, it is easy to understand the concerns about using wild type movement proteins as resistance genes in transgenic plants.

Post transcriptional gene silencing (PTGS)

PTGS is a universal mechanism by which plants are able to systemically switch off expression of targeted genes via the reduction of steady state level of specific RNAs. It is believed that PTGS to have evolved as a mechanism by which plants protect themselves from viral infection by targeting virus transgenes. Plant DNA viruses have properties with the potential to initiate gene silencing in the nucleus and in the cytoplasm.

Coat-protein (CP) mediated resistance

Coat or capsid proteins protect the virus nucleic acid from degradation. CP is also important to several stages in virus infection, including acquisition and transmission of viruses by insect vectors, cell to cell and long distance spread in the host, and for some viruses, CP regulates one or more steps of virus replication. CP genes have shown to be effective in preventing or reducing infection and disease caused by homologous or closely related viruses. CP-mediated protection has been reported against several viruses such as tobacco mosaic virus, tomato mosaic virus, cucumber mosaic virus, alfalfa mosaic virus, potato virus X, Y and leaf roll and several others. The protocol for CP-mediated resistance is given below.

Transformation of Truncated coat protein gene

The restricted fragment from recombinant clones are to be again ligated in a binary vector under the expression control of 35 S promoter using the same strategy as mentioned in chapter 24. Later it was mobilized to *Agrobacterium tumefaciens* using freeze-thaw method.

Cloning of coat protein gene into plant transformation vector

The *A. tumefaciens* Ti plasmid derived binary vector P^{B1121} may be used for making constructs containing CP gene. From the original vector, B-glucaronidase gene (gus) is removed leaving intact 35S promoter and NOS-terminater by restricting the vector with double digestion using Bam H I and Sac1. Sac 1 is used in place of Sst 1 since the sequence recognition is same for both the enzymes (Isozysomers). Corrosponding BamH I enzyme site is generated by oligonucleotide-directed mutagenesis in the forward primer and starting codon ATG are incorporated as well in a few nucleotides upstream to the N terminal end of CP gene. In the C-terminal site, the Sac I site used is the Sac I site of $P^{GEM\text{-}T}$ easy vector.

Preparation of vector and insert DNA fragment

Both CP gene and $P^{PBI\ 121}$ are prepared by extraction the plasmid from recombinant *E. coli*, and are restricted with BamHI and Sac I and checked on 1% agarose gel with 1X TAE buffer and the required products are purified by gel extraction kit (described earlier in chapter 24), and then used for ligation.

Transforming bacterial cells with ligation mix

The competent cells *E. coli* DH 5a for transformation are prepared by calcium chloride method. The eluted product is ligated with linearised $P^{PBI\ 121}$ as indicated in Table 9. This resulted in the ligation of CP gene in the sense direction to create coat protein sense. Later it was renamed as coat protein sense truncation (CpSTR).

Table 9. Recipe for ligation mix

Reagents	Volume required (µl per reaction)
Linearised $P^{PBI\ 121}$ vector	10
Insert	15
T_4DNA ligase	2
10 x ligase buffer	13
Total	**40**

Ligation mix is kept at 16°C in a water bath overnight. Before ligating coat protein gene, the vector DNA restricted ends are dephorylated with calf intestine alkaline phosphatase (CIAP @ 2 units/restriction mix) and incubated for 30 min at 37°C. The enzyme is then heat killed at 65°C for 2 h. After ligation, 200 µl competent cells of *E. coli* (strain DH 5a) are transformed with ligation mix. The cells are plated on to LA plates containing kanamycin (50 µg/ml) and incubated at 37°C.

The insert is confirmed by colony PCR from the transformed colonies. On the other hand, DNA is isolated from the transformed colonies and restricted with same enzymes used for cloning and analysis on 1% agarose gel in 1 x TAE buffer and further confirmed by southern hybridization with the probe made from PCR product as given below.

Southern hybridization

The same colony PCR product analysed in 1% agarose gel electrophoresis, is used for further confirmation by Southern hybridization. A ^{32}P labeled probe prepared from CP gene is used for confirmation. The procedure to be followed for Southern hybridization is given below:

1. Agarose gel after visualizing on UV transilluminator is cut at the position of the wells at the top and the RNAs at the bottom and measured accurately.
2. The top right corners are cut to fix the correct orientation of the gel.
3. The DNA in the gel is denatured by alkali by soaking it for 45 min in several volumes of denaturation solution (1.5 M NaCl, 0.5 N NaOH) (also see Appendix-1) with constant shaking.

4. The gel is then neutralized by soaking it in several volumes of neutralization solution (1 M Tris-HCl, pH 7.4, 1.5M NaCl, also see Appendix-1) for 45 min with constant shaking.
5. Finally, the gel is set up for capillary transfer of DNA onto NCM. NCM should be of the same size as is the gel and is also cut at the top right corner to fix the correct orientation of the gel.
6. 20 x SSC (NaCl sodium citrate buffer, Appendix-I) is used as transfer buffer.
7. The capillary transfer of DNA is allowed to proceed for 24 h.
8. The membrane is baked at 80°C for 2 h in vacuum.

The protocol for pre-hybridization, probe production, hybridization and autoradiography is the same as described in nucleic acid hybridization chapter

Plasmid induction into Agrobacterium (Triparental mating)

Cp gene construct in $P^{PBI\ 121}$ binary vector (maintained in *E. coli* strain DH 5a) (donor strain) is mobilized into disarmed *Agrobacterium tumefaciens* strain LBA 4404 by triparental mating procedure using the helper plasmid P^{RK2013} in *E. coli* /HB101 (helper strain).

In 50 ml LB medium containing kanamycin (50 µg/ml) (Appendix III), single colony of donor strain and helper strain in *E. coli* are inoculated for overnight growth separately at 37°C with continuous shaking at 200 rpm. *A. tumefaciens* strain LBA 4404 is inoculated in 50 ml LB medium having Rifampicin (50 µg/ml) at 28°C with continuous shaking at 200 rpm. This step must be taken one day before growing donor and helper. Next day 1 ml each of donor and helper, and 2 ml of recipient cell cultures are added in a sterile culture tube and mixed the three cultures together. From the 100 µl mixed cell suspension is placed onto the sterile nylone membrane (1 x 1 cm square) positioned in the centre of LA plate without any antibiotics. The plates are incubated at 28°C overnight for conjugation. After 16-18 h, the membrane with grown cells is placed in a beaker with 5 ml sterile distilled water. The cells are removed from the membrane by vigorous shaking and 100 µl cell suspension is spread on LA plates containing Rifampivin (50 µg/ml) and kanamycin (50 µg/ml) and incubated for 2 days at 28°C for appearance of single colonies of transconjugates. Later, nicely isolated single transconjugant colony is picked up and transferred to fresh LA plates having rifampicin and kanamycin antibodies and maintained at 28°C.

Confirmation of gene construct in Agrobacterium

Successful mobilization of gene construct in to *A. tumefaciens* strain LBA 4404 is confirmed by colony PCR.

Colony PCR

A single transconjugate colony of *A. tumefaciens* is picked up and rinsed with 20 µl of sterile distilled water (SDW) in an eppendorf tube. In order to break the bacterial cells and release the plasmid construct, the rinsed bacterial suspension is alternatively immersed in liquid nitrogen and boiling water. The suspension is then pulse centrifuged to remove debris and supernatant solution (containing plasmid) is used as template for PCR reaction.

Procedure to be followed thereafter is the same as DNA PCR described earlier. The colony PCR product is analysed in 1% agarose gel electrophoresis (Fig. 89).

TRANSFORMATION WITH COAT PROTEIN GENE

Source of explant

Seeds of the targeted plant species is sown in Murashige and Scoog medium, solidified with 1% agar. The pH of medium is set to 5.7. 4-5 seeds per majenta box are sown one to two weeks before transformation. The hypocotyls of young seedlings are excised to segment of 1-2 cm , each segment is cut with three open wound and used as explants for co-cultivation with Agrobacterium cells having coat protein gene construct.

Preparation of *A. tumefaciens* bacterial suspension for transformation

Single colony of *A. tumefaciens* containing the construct is transferred to 50 ml LB medium with rifampicin (50 µl/ml) and kanamycin (50 µg/ml). It is then incubated at 28°C for 36 h with shaking at 200 rpm. Bacterial cells without vector are also inoculated to serve as control. Well grown bacterial cells are palletized by centrifugation at 5000 rpm for 10 min and the pellet is suspended in 10 ml half strength modified MS medium without hormone.

Co-cultivation

Three days co-cultivation of explants is carried out with Agrobacterium having coat protein gene construct and kanamycin resistant selectable marker. The stem segments from the explant are submerged in *A. tumefaciens* culture for 10 min by gentle shaking to ensure that the bacteria are attached to all cut edges. Remove the segments and blotted dry them on sterile filter paper before transferring to the co-cultivation medium (Appendix III) and incubated in culture room with temperature of 26°C for two days. The explants submerged in half strength MS nutrient solution and co-cultivated with wild *Agrobacterium* (without construct) may serve as control. The segment explant co-cultivated with *A. tumefaciens* containing plasmid construct are simultaneously transferred to selection medium

(Appendix III). In each plate 15-20 co-cultivated leaf explants are transferred and the plates are kept under ambient growth conditions. The segments co-cultivated developed root and callus on co-cultivated medium.

Growth of transformed material

The shoot buds regenerated from transformed cut end of segments and control plants are excised and transferred to majenta box. After 2-3 weeks when the shoot buds elongated into plantlets, they are transferred into rooting medium (Appendix III). When profuse roots are produced in the rooting medium, plants are carefully pulled out of medium and adhering agar pieces are removed and they are acclimatized by keeping in MS salts devoid of sucrose for one week and in SDW for another week.

Genome DNA analysis for the transgene integration

The leaves of transformed plants were used for DNA extraction and following protocol as per manufacturer's instruction is followed.

1. Ground 100 mg leaves with liquid nitrogen to fine powder using a mortar and pestle, transfer tissue powder to an eppendorf.
2. Add 400 µl of buffer AP1 and 4 µl of RNase A stock solution (100 mg/ml) to a maximum of 100 mg of ground (wet weight) and vortex vigorously.
3. Incubate the mixture for 10 min at 65°C, mixed 2-3 times during incubation by inverting tune.
4. Add 130 µl of buffer AP2 to lysate, mixed and incubate for 5 min on ice.
5. Apply the lysate to the QIA shredder spin column sitting in a 2 ml collection tube and centrifuge for 2 min at maximum speed (table top centrifuge).
6. Transfer flow through fraction from step 5 to a new tube without disturbing the cell debris pellet.
7. Add 1.5 volume of buffer AP3 to the clear lysate and mix by pipetting.
8. Apply 650 µl of the mixture from step 7, including any precipitate which may have formed to the DNeasy mini spin column sitting in a 2 ml collection tube. Centrifuge for one min at 8000 rpm and discard flow through.
9. Place DNeasy column in a new 2 ml collection tube, add 500 µl buffer AW to the DNeasy column and centrifuge for 1 min at 8000 rpm. Discard flow through and reuse the collection tube in next step.
10. Add 500 µl buffer AW to the DNeasy column and centrifuge for 2 min at maximum speed to dry the membrane.

11. Transfer the DNeasy column to a 1.5 ml or 2ml microcentrifuge tube and pipette 100 µl of preheated (65°C) buffer AE directly onto the DNeasy membrane. Incubate for 5 min at room temperature and then centrifuge for 1 min at 8000 rpm to elute. It is ready for use in PCR.

The PCR is done with the same primers for checking the presence of transgene, then analyse in 1% agarose gel in 1 x TAE buffer and further confirmed by Southern hybridization with the probe made from PCR product (Fig. 90). The protocol has already been described earlier.

Expression of transformed coat protein gene

To check the expression of coat protein gene presence in the plant system, total protein from expected transformed plants is extracted by using PST buffer and further tested by DIBA using homologous antiserum as per the protocol described earlier.

CHAPTER 14

TAXONOMY OF VIRUSES — AN INTRODUCTION

Earlier virus classification was mainly host-oriented *i.e.* a virus was named on the basis of host plant it was found and the disease symptoms it caused. For example, the virus inducing mosaic symptoms in tobacco was called as "tobacco mosaic virus" or virus causing bunchy top symptoms in banana was named as "banana bunchy top virus". Johanson in 1927 suggested that a virus should be named by adding the word virus and a number to the common name for the host in which it was found. Thus tobacco mosaic virus was named as Tobacco virus 1.

Later Smith 1937 proposed that the name should be Latinized and the generic name of the host in which they were first found. Thus tobacco mosaic virus should be "*Nicotiana* virus 1". Further attempts were made by Holmes (1939, 1948) to use Latin Binomial system, under which tobacco mosaic virus should be called as "*Marmor tabaci*".

Gibbs *et al.* (1968) suggested that vernacular names should be retained and could be accompanied by a cryptogram summarizing the salient features of the virus in the code form (VAC system) For example:

TMV = R/1 : 2/5 : E/E : 5/0

I II III IV

I = Type of nucleic acid/No. of NA strands

R = RNA 1 = Single (D = DNA 2 = double)

II = Mol. Weight of NA in million / % Nucleic acid

III = Outline of particle shape / outline of nuclear capsid

S = Spherical

B = Bacilliform

E = Elongated

IV = Type of host infested / type of vector

B = Bacterium	Ap = aphids
F = Fungus	Au = Leafhopper
I = Invertebrate	Cl = Beetle
S = Seed plant (angiosperm)	Fu = Fungus
	Se = Seed transmitted,
	Ne = Nematode
	W = Whitefly, Th = Thrips
	0 = No vector

At the International Congress of Microbiology held in Reo de Janeiro in 1959, a subcommittee on virus nomenclature was formed and later an International Committee for the nomenclature of virus was established and finally, International Committee on Virus Taxonomy (ICTV) was established with the objectives:

1 To develop an internationally agreed taxonomy for viruses.
2. To establish internationally agreed names for taxonomic groups of viruses.
3. To communicate latest results on the classification and nomenclature of viruses to virologists by holding meetings and publishing results.

Two major events have made a remarkable impact on plant virus taxonomy after 1993.

1. The acceptance by PVS of ICTV and the whole virology committee of the family-genus-species classification system.
2. The advent of the molecular era leading to "phylogenetic tree" and "quantitative taxonomy". The most recent informations have been used to classify viruses based on nucleotide and amino acid sequences. It is becoming more and more prevalent as exemplified by the significant number of phylogenetic trees in VII ICTV report.

   It was only in 1995, that in its VI th report ICTV proposed a uniform system for all viruses. Viruses are classified into order, family, genera and species. There were some unassigned (floating) genera. Virus taxa in VII report were arranged on the basis of, nature of virus genome, strandedness of virus genome, polarity of virus genome and facility for reverse transcription. Tables 10 and 11 give the reports on taxonomy of viruses.

Table 10. The Current System of Plant Virus Classification

Order	Family	Sub-family	Genus	Type species	Host
The DNA virus : the ds DNA viruses					
	Adenovirus		Rhizidiovirus	Rhizidiomyces virus	Fungi
	the DNA viruses				
	Geminiviridae		Master virus	Maize streak virus	Plants
			Curtovirus	Beet curly top virus	Plants
			Begomovirus	Been golden mosaic virus	Plants
			Topovirus	Tomato pseudo-curly top virus	Plants
	Cirocoviridae		Nanovirus	Subterranean clover stunt virus	Plants
The DNA and RNA reverse transcribing viruses					
	Caulimoviridae		Caulimovirus	Cauliflower mosaic virus	Plants
			PVCV-like viruses	Petunia vein-clearing virus	Plants
			SbCMV-like virus	Soybean chlorotic mottle virus	Plants
			CsVMV-like virus	Cassava vein mosaic virus	Plants
			Badnavirus	Commelina yellow mottle virus	Plants
			RTBV-like virus	Rice tungro bacilliform virus	Plants
The RNA viruses : the dsRNA viruses					
	Reoviridae		Fijivirus	Fiji disease virus	Plants
			Phytoreovirus	Wound tumor virus	Plants
			Oryzavirus	Rice ragged stunt virus	Plants
	Totiviridae		Totivirus	Saccharomyces Cerevisiae virus L-A	Fungi

Table 10. (*Contd.*)

Order	Family	Sub-family	Genus	Type species	Host
	Partitiviridae		Partitivirus	*Gaeum-annomyces graminis* virus 0196/6-A	Fungi
			Chrysovirus	*Penicillium chrysogenum* virus	Fungi
			Alphacrypto virus	White clover cryptic virus I	Fungi
			Betacrypto virus	White clover cryptic virus II	Fungi
	Hypoviridae		Hypovirus	Cryphonectria hypovirus I EP713	Fungi
			Varicosavirus	Lettuce big-vein virus	Plants
The negative-stranded ssRNA viruses					
Mononegavirales	Rhabdoviridae		Cytohabdo virus	Lettuce necrotic yellow virus	Plants
			Nucleorhabdo virus	Potato yellow dwarf virus	Plants
	Bunyaviridae		Tospovirus	Tomato spotted wilt virus	Plants
			Tenuivirus	Rice stripe virus	Plants
			Ophiovirus	*Citrus psorosis* virus	Plants
The positive-stranded ssRNA viruses					
	Sequiviridae		Sequivirus	Parsnip yellow fleck virus	Plants
			Walkavirus	Rice tungro spherical virus	Plants
	Comoviridae		Comovirus	Cowpea mosaic virus	Plants
			Fabavirus	Broad bean wilt virus	Plants
			Nepovirus	Tobacco ringspot virus	Plants
	Potyviridae		Potyvirus	Potato virus Y	Plants
			Rymovirus	Ryegrass mosaic virus	Plants
			Macluravirus	Maclura mosaic virus	Plants

Table 10. (*Contd.*)

Order	Family	Sub-family	Genus	Type species	Host
			Ipomovirus	Sweet potato mild mottle virus	Plants
			Bymovirus	Barley yellow mosaic virus	Plants
			Tritimovirus	Wheat streak mosaic virus	Plants
			Sobemovirus	Southern bean mosaic virus	Plants
			Marafivirus	Maize rayado fino virus	Plants
	Luteoviridae		Luteovirus	Barley yellow dwarf virus-MAV	Plants
			Polerovirus	Potato leafroll virus	Plants
			Enamovirus	Pea enation mosaic virus	Plants
			Umbravirus	Carrot mottle virus	Plants
	Tombus viridae		Avenavirus	Oat chlorotic stunt virus	Plants
			Aureusvirus	Pothos latent virus	Plants
			Carmovirus	Carnation mottle virus	Plants
			Dianthovirus	Carnation ringspot virus	Plants
			Machlomo virus	Maize chlorotic mottle virus	Plants
			Necrovirus	Tobacco necrosis virus	Plants
			Panicovirus	Panicum mosaic virus	Plants
			Tombusvirus	Tomato bushy stunt virus	Plants
			Tobamovirus	Tobacco mosaic virus	Plants
			Tobravirus	Tobacco rattle virus	Plants
			Hordeivirus	Barley strip mosaic virus	Plants
			Furovirus	Soil-borne wheat mosaic virus	Plants

Table 10. (*Contd.*)

Order	Family	Sub-family	Genus	Type species	Host
			Pomovirus	Potato mop-top virus	Plants
			Pecluvirus	Peanut clump virus	Plants
			Benyvirus	Beet necrotic yellow vein virus	Plants
	Bromoviridae		Alfamovirus	Alfaalfa mosaic virus	Plants
			Bromovirus	Brome mosaic virus	Plants
			Cucumovirus	Cucumber mosaic virus	Plants
			Ilarvirus	Tobacco streak virus	Plants
			Oleavirus	Olive latent virus 2	Plants
			Ourmiavirus	Ourmia melon virus	Plants
			Ideaovirus	Raspberry bushy virus	Plants
	Closteroviridae		Closterovirus	Beet yellow virus	Plants
			Crinivirus	Lettuce infectious yellow virus	Plants
			Capillovirus	Apple stem grooving virus	Plants
			Trichovirus	Apple chlorotic leaf spot virus	Plants
			Vitivirus	Grapevine virus A	Plants
			Tymovirus	Turnip yellow mosaic virus	Plants
			Carlavirus	Carnation latent virus	Plants
			Potexvirus	Potato virus X	Plants
			Allexivirus	Shallot virus X	Plants
			Foveavirus	Apple stem pitting virus	Plants
	Barnaviridae		Barnavirus	Mushroom bacilliform virus	Fungi

Unassigned viruses

Sub viral agents: viriods, satellites and agents of spongiform encephalopathies (peions)

Sub viral agent	Family	Genus	Type species	Host
Viriods	Pospiviroidae	Pospiviroid	Potato spindle tuber viriods	Plants
		Hostuviroid	Hop stunt virus	Plants
		Cocadviriod	Coconut cadang-cadang viroid	Plants
		Apscaviriod	Apple scar skin viroid	Plants
		Coleviroid	*Coleus blumei* viroid	Plants
	Avsunviroidae	Avsunviroid	Avocade sunblotch viroid	Plants
		Pelamoviroid	Peach latent mosaic virus	Plants
Satellite				Plants Inver-tebrates
				Fungi
Prions				Verte-brates
				Fungi

Source: Fauquet, 1999

In 2004, ICTV approved a proposal to create a plant virus family Flexiviridae. The family is named because its members have flexuous virions and it includes the existing genera, Allexivirus, Capillovirus, Carlavirus, Foveavirus, Potexvirus, Trichovitus and Vitivirus, plus the new genus Mandrivirus and some related viruses not assigned to any genus. The family is justified from phylogenetic analysis of the polymerase and coat protein (CP) sequences. To help to define suitable criteria for demarcation of species, a complete set of pair-wise comparison was made using the nucleotide (nt) and amino acid (aa) sequences of each fully sequenced gene from every available accession in the family. Based on the distributions and on inspection of data, it was concluded that, as a general rule, less than ca. 72% identical nt or 80% identical aa between their entire CP or replication protein gene.

In 2005, in 8th report of ICTV, the taxonomic structure of virus orders, families and genera are listed in three different ways to facilitate the search for a particular virus family or genus.

Order, family and genus name are sorted

1. In alphabetical order

Table 11. Order, families and floating genera of viruses according to the seventh ICTV report (1999)

Criteria	Order	Family	Floating genus	Morphology	Genome configuration	Genome size (Kb)	Virus host	Specie	Strain/ serotype	Tenta-tive	Total
ds DNA			Rhizidiovirus	Isometric	1 linear	27	Fungus	1	0	0	1
ss DNA		Geminividae		Isometric	1 or 2 circular	3-6	Plant	94	2	10	106
		Circoviridae		Isometric	1 circular	1.7-2.3	Vertebrate	3	0	1	4
			Nanovirus	Isometric	6-9 circular	6-9	Plant	4	0	1	5
ds RNA RT		Caulimovi-ridae		Isometric bacilliform	1 circular	8	Plant	26	0	8	34
ss RNA sRT		Metavirus		Isometric	1 linear	4-10	Yeast,fungus, invertebrate	18	0	1	19
ds RNA		Reoviridae		Isometric	10-12 segments	19-62	Vertebrate, invertebrate, plants	62	256	39	357
		Totiviridea		Isometric	1 segment	5-7	Fungus-protozoa	18	0	5	23
		Partiviridae		Isometric	2 segments	3-10	Fungus, plants	30	0	15	45
		Hypoviridae		Pleomorphic	1 segment	9-13	Fungus	3	0	2	5
			Varicosavirus	Rod	2 segments	14	Plants	1	0	3	4
Negative ssRNA	Mono-negav-irales	Rhabdoviridae unyaviridae		Bacilliform	1 segment	10-13	Vertebrate, plants	37	0	142	179
			Tospovirus	Spherical	3 segments	12-23	Vertebrate, plants	93	236	66	395
			Tenuivirus	Filaments	4-segments	15-19	Plants	6	0	5	11
			Ophiovirus	Filaments	3 segments	12	Vertebrate	3	0	0	3
Positive ssRNA		Equiviridae		Isometric	1+ segment	9-12	Plants	5	0	0	5
		Comoviridae		Isometric	2+ segments	9-16	Plants	50	0	9	59
		Potyviridae		Rod	1 or 2 + segments	8-12	Plants	106	0	92	198

Table 11. (*Contd.*)

Criteria	Order	Family	Floating genus	Morphology	Genome configuration	Genome size (Kb)	Virus host	Specie	Strain/ serotype	Tenta- tive	Total
			Sobemovirus	Isometric	1 + segment	4	Plants	11	0	3	14
			Marafivirus	Isometric	1 + segment	6-7	Plants	3	0	0	3
		Luteoviridae		Isometric	1 or 2 + segments	6-9	Plants	8	0	11	19
			Umbravirus	No particle	1 + segment	4	Plants	7	0	15	22
		Tombusviride		Isometric	1 or 2 + segments	4-5.5	Plants	38	0	11	49
			Fovaevirus	Rod	1 + segment	8-9	Plants	2	0	1	3
ssRNA Positive sense			Tobamovirus	Rod	1 + segment	6	Plants	16	0	3	19
			Tobravirus	Rod	2 + segment	9-11	Plants	3	0	0	3
			Hordevirus	Rod	3 + segment	10	Plants	4	0	0	4
			Furovirus	Rod	2 + segment	9-11	Plants	1	0	4	5
			Pomovirus	Rod	3 + segment	12	Plants	4	0	0	4
			Pecluvirus	Rod	2 + segment	10	Plants	2	0	0	2
			Benyvirus	Rod	4 (or 5) + segment	14-16	Plants	2	0	0	2
		Bromoviridae		Isometric, bacilliform	3 + segment	8-9	Plants	28	0	0	28
			Ourmiavirus	Bacilliform	3 + segment	4-5	Plants	3	0	0	3
			Idaeovirus		3 + segment	8	Plants	1	0	0	1
		Clostero-viridae		Rod	1 or 2 + segment	15-19	Plants	18	0	16	34
			Capillovirus	Rod	1 + segment	7	Plants	3	0	1	4
			Trichovirus	Rod	1 + segment	7.5	Plants	3	0	1	4
			Vitivirus	Rod	1 + segment	7.5	Plants	4	0	1	5

Table 11. (*Contd.*)

Criteria	Order	Family	Floating genus	Morphology	Genome configuration	Genome size (Kb)	Virus host	Specie	Strain/ serotype	Tenta- tive	Total
			Tymovirus	Isometric	1 + segment	6	Plants	20	0	1	21
			Carlavirus	Rod	1 + segment	7-8	Plants	31	0	29	60
			Potevirus	Rod	1 + segment	6	Plants	26	0	18	44
			Alleivirus	Rod	1 + segment	9	Plants	6	0	3	9
			Fovaevirus	Rod	1 + segment	8-9	Plants	2	0	1	3
		Barmavi-radae		Bacilliform	1+ segment	4	Fungus	1	0	1	2
		Unassigned virus		–	–	–	All	30	0	0	30
		Viroids		–	–	–	Plant	27	0	8	35
		Setellites		–	–	–	Plant	33	0	6	39

Source: Fauquet 1999

2. According to their taxonomic order based on nucleic acid content and strandedness
3. According to their host range

Throughout the document taxa are listed to their taxonomic order which is determined by phenotypic and genotypic properties. Each type species name is preceded by the name of the order, family and genus it belongs to. Names of orders, families, and genera are in italic script, if they are approved by ICTV. Taxa names in quotes (and not in italic script) indicate that this taxon has not an international ICTV approved name. Species (vernacular) names are since 1999 also given in italic script. The detail classification system was proposed by Fauquet *et al.*, 2005 (http://ncbi.nlm.nih.gov/ICTVdb/Ictv/fr-fst-g.htm).

The details of 7th ICTV report is given in Tables 10 and 11. In the recent taxonomy document taxa are listed according to their taxonomic order. Each type species name is preceded by the name of the order, family and genus it belongs to. Name of orders, families and genera are in italic script. For details see Fauquet *et al.*, 2005.

Some useful information

These are given at the end of this guide relating to preparation of various buffers, reagents (Appendix I-III) and conversion tables for the use of researchers and students.

SELECTED READING

Ahlawat, Y.S., Ramachandran, P., Malathi, V.G. and Verma, J.P. 1996. Detection of plant pathogens. A publication of the Centre for Advanced Studies in Plant Pathology, Indian Agricultural Research Institute, New Delhi. pp. 279.

Bos, L. 1983. Introduction to plant virology. Oxford & IBH Publishing Co. New Delhi, pp. 160.

Christie, R.G. and Edwardson, J.R. 1997. Light and electron microscopy of plant virus inclusions. Florida Experiment Station Monograph Seriese No.9, pp. 150.

Dijkstra, J. and de Jager, C.P. 1998. Practical plant virology: protocols and excercises, Springer, New York, pp. 459.

Ehrlich, H.A (*ed*) 1989. PCR technology - Principles and application for DNA amplification. Stockton Press, New York, pp. 246.

Griffith, O.M. 1979. Technique of preparative, zonal and continuous flow ultracentrifugation. pp. 1-50, Beckman Instruments Inc.

Hampton, R., Ball, E. and de Boer, S. 1990. Serological methods for detection and identification of viral and bacterial plant pathogens. A laboratory manual. APS press, St. Paul, MN.

Harris, K.F. and Maramorosch, K. 1977. Aphids as virus vectors. Academic press, New York, pp. 559.

Harris, K.F. and Maramorosch, K. 1980. Vector of plant pathogens. Academic press, New York, pp. 467.

Holmes, F.O. 1964. Symptomatology of virus diseases in plants. *In*: M.K. Corbett and H.D. Sisler (*ed*): Plant virology. University of Florida Press, Gainesville, pp.17-38.

Jones, R.A.C. and Torrance, L. 1986. Development and application in virus testing. Lavenham Press, Suffolk, UK, pp. 300.

Kado, C.I. and Agarwal, H.O. 1976. Principle and techniques in plant virology, Van Nostrand, Reinhold, New York.

Kurstak, E. 1981. Handbook of plant virus infections. Comparative diagnosis Elsevier/ North Holland, Amsterdam, pp. 943.

Mandahar, C.L. (*ed*) Plant Viruses Vol. II Pathology. CRC Press, Boca Raton, FL. USA.

Maramorosch, K. and Harris, K.F. (*eds*) 1979. Leafhopper vectors and plant disease agents. Academic press, New York, pp. 464.

Maramorosch, K. and Koprowski, H. 1972. Methods in Virology, Vol. IV, Academic press, New York.

Matthews, R.E.F. 1991. Plant virology 3rd edn. Academic press, London.

Noordam, D.1973. Identification of plant viruses; Methods and Experiments. Pudoc. Wageningen, pp. 207.

Ouchterlony, O. 1968. Handbook of immunodiffusion and immunoelectrophoresis. Ann Arbor Scientific Publications, Michigan.

Raychaudhuri, S.P. and Nariani, T.K. 1977. Virus and mycoplasma diseases of plants in India. Oxford & IBH Publishing Co. New Delhi, pp. 102.

Sambrook, J. and Russell, D.W. 2001. Molecular cloning - a laboratory manual. Cold spring Harbour Laboratory press, New York.

SELECTED REFERENCES

Adams, M.J., Antoniw, J.F., Bar-Joseph, M., Brunt, A.A., Candresse, I., Foster, G.D., Martelli, G.P., Milne, R.G. and Fauquat, C.M. 2004. The new plant virus family flexiviridae and assessment of molecular criteria for species demarcation. *Arch. Virol.* **149**: 1045-1060.

Aebig, J.A., Jordan, R.L., Lawson, R.H. and Hsu, H.T. 1987. Immunological and biological properties of a mouse monoclonal antibody reactive to prunus necrotic ringspot virus. *Intervirology* **28**: 57-64.

Ahlawat, Y.S. and Chenulu, V.V. 1984. Radish mosaic: a new disease caused by turnip mosaic virus in India. *Tropical Agriculture* **61**: 188-192.

Ahlawat, Y.S., Varma, A., Chakraborty, N.K. and Srivastava, K.P. 1995. Biotypes and vector capacity of *Diaphorina citri* in transmitting citrus greening BLO. *Proc. 2nd agricultural Sci. Cong*. pp. 222-232.

Baranwal, V.K., Majumdar, S., Ahlawat, Y.S. and Singh, R.P. 2003. Sodium sulphite yields improved DNA of higher stability for PCR detection of citrus yellow mosaic virus from citrus leaves. *J. Virol. Methods* **112**: 153-156.

Bennett, C.W. 1967. Plant viruses transmission by dodder. *In*: K. Maramorosch and H. Koprowski (*eds*): Methods in Virology Vol. I. Academic Press New York, pp. 403-410.

Bennett, C.W. 1969. Seed transmission of viruses. *Adv. Virus Research* **14**: 221-258.

Birnboim, H.C. and Doly, J. 1979. A rapid alkaline extraction procedure for screening recombinant plasmid DNA. *Nucl. Acid Res*. **7**: 1513-1523.

Bove, J.M., Garnier, M., Ahlawat, Y.S., Chakraborty, N.K. and Varma, A. 1993. Detection of Asian strains of greening BLO by DNA-DNA hybridization in Indian orchard trees and Malaysian *Diaphorina citri* psyllids. *Proc. 12th Conf. IOCV*, IOCV, Riverside, California.

Byadgi, A.S. and ahlawat, Y.S. 1995. A new viral ringspot disease of citrus (*Citrus* species) in India. *J. agric. Sci.* **65**: 61-68.

Campbell, R.N. 1979. Fungal vectors of plant viruses. *In*: H.P. Molitoris, M. Hollings and H.A. Wood (*eds*) Fungal Viruses. Springer verlag, Berlin, pp. 8-24.

Clark, M.F. and Adams, A.N. 1977. Characteristics of the microplate method of enzyme-linked immunosorbent assay for the detection of plant viruses. *J. Gen. Virol*. **34**: 475-483.

Comoi, L. schilling-Cordovo, C., Mergia, A. and Houck, C.M. 1983. A new technique for genetic engineering of *Agrobacterium tumefaciens* Ti plasmid. *Plasmid* **10**: 21-30.

Derrick, K.S. 1973. Quantitative assay for plant viruses using serologically specific electron microscopy. *Virology* **56**: 652-653.

Derrick, K.S. and Brlansky, R.H. 1976. Assay for viruses and mycoplasma using serologically specific electron microscopy. *Phytopathology* **66**: 815-820.

Diener, T.O. 1981. Viroids. *Scientific American* **244**: 58-65.

Doi,Y., Terenaka, M., Yora, K. and Asuyama, H. 1967. Mycoplasma or PLT group-like microorganisms found in the phloem elements of plants infected with mulberry dwarf, potato witch's broom, aster yellows or Paulownia witch's broom. *Annals of the Phytopathological Society of Japan* **33**: 259-266.

Fauquet, C.M. 1999. Encyclopedia of virology. Academic Press New York, pp. 1730-1756.

Fauquet, C.M., Mayo, M.A., Maniloff, J., Desselberger, U. and Ball, L.A. (eds). 2005. Virus Taxonomy: 8th report of the International committee on Taxonomy of Viruses. Academic Press, 1162 pp. Elsevier.

Feinbezg, A.P. and Vogelstein, J. 1984. A technique for radio-labeling DNA restriction endonucleases fragments to high specific activity. *Ann. Biochem.* **137**: 266-67.

Genga, A.M. 1991. Genetic transformation in *Phaseolus* species by high velocity microprojectiles. *Acta Hort* **300**: 309-313.

Gerlach, W.L., Liwellyn, D. and Haselloff, J. 1987. Construction of a plant disease resistance gene from satellite RNA of tobacco ringspot virus. *Nature* **328**: 802-805.

Ginoza, W. and Atkinson, D.E. 1955. Comparison of some physical and chemical properties of eight strains of tobacco mosaic virus. *Virol.* **1**: 253-260.

Golemboski, D.B., Lomonssoff, G.P. and Zaitlin, M. 1990. Plants transformation with a tobacco mosaic virus non-structural gene sequences are resistant to the virus. *Proc. Natl. Acad. Sci. USA* **87**: 6311-6315.

Golles, R., Moser, R., Puhringer, H., Katinger, H., laimer da Camara and Machado, M. 2000. Transgenic grapevines expressing coat protein gene sequences of grapevine virus A and grapevine virus B. *Acta Hort.* **528**: 305-311.

Hammond, J. and Kamo, K.K. 1993. Transgenic coat protein and antisense RNA resistant to bean yellow mosaic potyvirus. *Acta Hort.* **336**: 171-181.

Haq, Q.M.R., Srivastava, K.M., Raizada, R.K., Singh, B.P., Jain, R.K., Mishra, A. and Shukla, D.D. 1994. Biological, serological and coat protein properties of a strain of turnip mosaic virus causing a mosaic disease of *Brassica campestris* and *B. juncea* in India. *J. Phytopath.* **140**: 55-64.

Harris, K.F. 1972. Arthropod and Nematode vectors of plant Viruses. *Ann. Rev. Phytopathol.* **19**: 391-426.

Hebert, T.T. 1963. Purification of Plant viruses by polyethylene glycol. *Phyt.* pp. 53-362.

Hoa, N.V. and Ahlawat, Y.S. 2004. Characterization of four isolates of Indian Citrus Ring spot Virus. *Indian Phytopath.* **57**: 296-302.

Hoa, N.V., Ahlawat, Y.S. and Pant, R.P. 2004. Production of virus-free Kinnow mandarin and Mosambi sweetorange nucleus planting material through shoot tip grafting. *Indian Phytopath.* **57**: 482-487.

Hobbs, H.A., Reddy, D.V.R., Rajeshwari, R. and Reddy, A.S. 1987. Use of direct antigen coating method and protein-A coating ELISA procedures for detection of three peanut viruses. *Plant Dis.* **71**: 315-320.

Hofgen, R. and Willmitza, L. 1988. Storage of competent cells of *Agrobacterium tumefaciens. Nucl. Acid Res.* **16**: 9877.

Horsch, R.B., Fry, J.E., Hoffmann, N.L., Eicholtz, D., Rogers, S.G. and Fraley, R.T. 1985. A simple and general method for transferring genes to plants. *Science* **277**: 1229-1231.

Keonig, R. and Paul, H.L. 1983. Detection and differentiation of plant viruses by various ELISA procedures. *Acta Hort.* **127**: 147-158.

Koenig, R. and Paul, H.L. 1982. Variants of ELISA in plant virus diagnosis. *J. Virol. Methods* **5**: 113-125.

Mandahar, C.L. 1999. Molecular biology of transgenic plants. pp. 241-269. *In*: Molecular biology of pant viruses. *ed.* Mandahar, C.L.

Manoussopoulos, I.N., Maiss, E. and Tsagris, M. 2000. Native electrophoresis and Western blotting analysis (NEWeB): a method for characterization of different forms of potyvirus particles and similar nucleoprotein complex in extracts of plant tissues. *J. Gen. Virol.* **81**: 2295-2298.

Maule, A.J., Hull, R. and Dawson, J. 1983. The application of spot hybridization to the detection of DNA and RNA viruses in plant tissues. *J. Virol Methods* **6**: 215-224.

McCornic. 1999. Transformation of tomato with *Agrobacterium tumefaciens. Plant Tissue Cult. Manual* B6, pp. 1-9.

Milne, R.G. and Luisoni, E. 1975. Rapid high resolution immune electron microscopy.

Moore, G.A. 1985. Factors affecting *in vitro* embryogenesis from underdeveloped ovules of mature citrus fruit. *American Soc. Hort. Sci.* **1**: 66-70.

Morris, T.J. and Dodds, J.A. 1979. Isolation and analysis of double stranded RNA from virus infected plant and fungal tissue. *Phyt.* **69**: 854-858.

Navaro, L. 1981. Shoot-tip grafting *in vitro* and its application: a review. *Proc. Int. Soc. Citriculture* **1**: 452-456.

Owens, R.A. and Diener, T.O. 1981. Sensitive and rapid diagnosis of potato spindle tuber viroid disease by nucleic acid hybridization. *Sci.* **213**: 670-671.

Palukaitis, P. and Zaitlin, M. 1997. Replicase-mediated resistance to plant virus diseases. *Adv. Virus Res.* **48**: 349-377.

Pappu, H.R., Niblett, C.L. and Lee, R.F. 1995. Application of recombinant DNA technology to plant protection. Molecular approaches to engineering virus resistance in crop plants. *World J. Microbial. Biotech.* **11**: 426-437.

Pawley, A.R. 1973. Year-round production of test plants for virology. Glasshouse Crop Res. Inst. Ann. Rept. pp. 149-151.

Ramachandran, P., Roy, A., Mathur, S., Agrawal, J., and Ahlawat, Y.S. 2003. Diagnosis of citrus exocortis and hop stunt viroids associated with yellow corky vein disease in citrus. *Indian Phytopath.* **56**: 428-433.

Sambade, A., Martin, S., Olmos, A., Garcia, M.L., Cambra, M., Grau, O., Guerri, Z. and Moreno, P. 2000. A fast one-step reverse transcription polymerase chain reaction (RT-PCR) amplification procedure providing highly specific complementary DNA from plant virus RNA. *J. Virol. Methods* **87**: 25-28.

Sambrook, J., Fritsch, E.F. and Maniatis, T. 1989. Molecular cloning - a laboratory manual. Cold Spring Harbour Laboratory, New York.

Shoujiang Gao, Monique Garnier and Joseph M. Bove. 1993. Production of monoclonal antibodies recognizing most Asian strains of the greening BLO by *in vitro* immunization with an antigenic protein purified from the BLO. *Proc. 12th Conf. IOCV,* IOCV, Riverside, California.

Singh, R.P. 2009. Multiplex reverse transcription polymerase chain reaction: A versatile technique for multiple virus and viroid detection. *In*: Virus and virus-like diseases of plants – IPM system in Agriculture Vol. X (*eds*) Upadhyay, R.K., Ahlawat, Y.S., Dube, O.P., Srivastava, K.P. and Baranwal, V.K. Aditya Books Pvt. Ltd., New Delhi.

Smith, F.D. and Banttari, E.E. 1987. Dot-ELISA on nitrocellulose membranes for detection of potato leaf roll virus. *Plant Dis.* **7**: 795-799.

Smith, K.M. 1972. A text book of plant virus diseases. Longmann: Harlow.

Stace-Smith, R. 1965. A simple apparatus for preparing sucrose density gradient tubes. *Phyt.* **55**: 1031.

Starrantino, A. and Caponnetto, P. 1990. Effect of cytokinins on embryogenic callus formation from developed ovules of orange. *Acta Hort.* **280**: 191-194.

Steere, R.L. 1959. The purification of plant viruses. *Adv. Virus Research* **6**: 1-73.

Teakle, D.S. 1983. Zoosporic fungi and viruses: double trouble. *In*: Zoosporic plant pathogens (*ed.* Buezacki, S.T.) Academic Press, New York , pp. 231-248.

Tomlinson, J.A. 1963. Effect of phosphate and borate on the infectivity of some viruses during purification. *Nature* **200**: 93-94.

Van Regenmortal, M.H.V. 1982. Serology and immunochemistry of plant viruses. Academic press, New York.

Vijaya Kumari, N. and Shyam Singh. 2000. *In vitro* micrografting and its applications in citrus industry. pp. 231-234. *In*: Hi tech citrus management, improvement, Biotechnology, production system, pest and disease management, postharvest technology (*eds*) Shyam singh and S.P.Ghosh, NRcC, Nagpur.

Walkey, D.G.A. 1985. Applied plant virology. William Heinemann Ltd. London.

Weir, D.M. (*ed*) 1978. Handbook of experimental immunology. Blackwell, Oxford.

Yarwood, C.E. and Fulton, R.W. 1967. Mechanical transmission of plant viruses. *In*: K. Maramorosch and H. Koprowski (*eds*): Methods in Virology Vol. I. Academic Press Newyork, pp. 237-266.

Yogita Verma, Shivali sood, Ahlawat, Y.S., Paulkhurana, S.M., Nie X, and Singh, R.P. 2003. Evaluation of multiplex reverse transcription polymerase chain reaction (RT-PCR) for simultaneous detection of potato viruses and strains. *Indian J. Biotech*. **2**: 587-590.

INTERNET SEARCH

An introduction of DNA Microarray 2009.
http://www.en.wikipedia.org/wik/phage_display
http://www.ncbi.nlm.gov./about/primer/microarrays.html.
http://www.pacificimmunology.com/typesofantibodies.asp
http://www.premierbiosoft.com/tech_notes/microarray.html
Microarrays Factsheet, National centre for Biotechnology information 2009.
Nicole Stricker and Min Li. 2001. Encyclopedia of life sciences/nature publishing group,
Phage display from Wikipedia, the free encyclopedia. 2009.
Types of /polyclonal, monoclonal antibody production/monospecific.
www.els.net.

APPENDIX I

1. Buffers for Mechanical Transmission and Purification

1.1. *Inoculation buffer (Phosphate buffer for 1L,0.01 M (pH 7.2)*

(i) Potassium dihydrogen phosphate (KH_2PO_4) – 1.362 g/1

(ii) Disodium hydrogen phosphate dehydrate ($NaH_2PO_4.2H_2O$) – 1.781 g/1

Mixed 285 ml of solution (i) with 715 ml solution (ii) gives 1 litre of phosphate buffer pH 7.2 and 0.01 M

1.2. *Pufication buffer (Phosphate buffer for 1L,0.05 M (pH 7.8)*

(iii) Potassium dihydrogen phosphate (KH_2PO_4) – 6.810 g/1

(iv) Disodium hydrogen phosphate dehydrate ($NaH_2PO_4.2H_2O$) – 8.905 g/1

Mixed 86 ml of solution (i) with 914 solution (ii) gives 1 litre of phosphate buffer pH 7.8 and 0.05 M

2. Enzyme-Link Immunosorbent Assay (ELISA)

(i). *Phosphate buffer saline (PBS) pH 7.4 (1 X)*

NaCl	– 8.0 g
KH_2PO_4	– 0.2 g
$Na_2HPO_4.2H_2O$	– 2.9 g
KCl	– 0.2 g
NaN_3	– 0.2 g
Distilled water	– 1000 ml

(ii) Coating buffer (pH 9.2)

Na_2CO_3	– 1.59 g
$NaHCO_3$	– 2.93 g
NaN_3	– 0.20 g
Distilled water	– 1000 ml

(iii) *Washing buffer (PBS-T)*

1 × PBS	– 100 ml
Tween 20	– 900 ml

(iv) *Antigen extraction buffer*

Coating buffer containing 2% solute polyvinyl pyrrolidone (MW 44,000)

(v) *Enzyme conjugate buffer*

PBS-T	– 1000 ml
Polyvinyl Pyrrodone (PVP)	– 20.0 g
Ovalbumin-40	– 2.0 g

(vi) *Blocking solution*

Add 5 g of Bovine Serum Albumin (BSA) to 1000 ml coating buffer

(vii) *Substrate buffer (pH 9.8)*

Diethanol amine	– 97 ml
Distilled water	– 800 ml
NaN_3	– 0.20 g

The pH adjusted to 9.8 with ^{1}N HCl and make upto 1000 ml with distilled water.

APPENDIX II

Common Reagents, Buffer and Media Used for Cloning and Confirmation of Clone

(i) Antibiotics

***Ampicillin*:** Stock solution (50 mg/ml) of the antibiotic was made made in distilled water, filter sterilized (through 0.22 micron filter) and distributed into 200 µl aliquots and stored at -20°C. It was used at a concentration of 50 µg/ml.

(ii) Electrophoresis reagents

50 × TAE	Tris base	242.0 g
	Glacial acetic acid	57.1 ml
	0.5 M EDTA (pH 8.0)	100.0 ml
	Distilled water to 1 litre.	
Loading dye	1% Bromophenol blue	200 µl
	Glycerol	200 µl
	10% SDS	60 µl
	10 × TAE	60 µl
	Distilled water	30 µl

(iii) Plasmid isolation buffers

Sol. I (resuspension buffer)	25 mM Tris HCl (pH 8.0)
	50 mM Glucose
	10 mM EDTA
Sol. II (lysis buffer)	0.2 N NaOH
	1% SDS
Sol. III (neutralization buffer)	3 M Sodium acetate pH 4.8

(iv) Hybridization solutions

Denaturing solution	1.5 M NaCl
	0.5 M NaOH

Neutralizing solution	1.5 M NaCl 1 M Tris. HCl (pH 7.4)
Denhardt's reagents (50×)	1 g Polyvinylpyrilidone (Sigma) 1 g Ficoli (Type 400, Sigma) 1 g Bovine serum albumin (Fraction V. Sigma) 100 ml Double distilled water.

This stock solution was stored at -20°C. It was used at working concentration of 5 x Denhardt's reagent.

Standard saline citrate (SSC) buffer 20 ×, pH 7.0

NaCl (3.0 M)	175.3 g
Sodium citrate (0.3 M)	88.2 g

Dissolved in 800 ml distilled water. Adjusted pH to 7.0 and made up the volume to 1 litre. Dispensed into aliquots and sterilized before storing at room temperature.

(v) Prehybridization solution

5 × Denhardt's reagent
6 × SSC
0.3% SDS
100 μg ml^{-1} yeast RNA

(vi) Other buffer

Membrane washing buffer	0.1% SDS 2 × SSC
Dephophorylation buffer	10 mM Mg Cl_2 100 mM tris. HCl (pH 8.3)
10× Ligation buffer	0.5 M tris. HCl (pH 7.6) 0.5 M Mg Cl_2 0.1 M Dithiothreitol 500 μg ml^{-1} BSA (Sigma Chemical Co., A 8022)

(vii) DNA molecular weight marker

1 Kilobase (1 Kb) DNA ladder of MBI Ferment was used as marker. The

Ladder formed by fourteen DNA fragments of 10 kb, 8 kb, 6 kb, 5 kb, 4 kb, 3 kb, 2.5 kb , 2 kb, 1.5 kb, 1 kb, 0.75 kb, 0.5 kb and 0.25 kb.

(viii) Preparation of commonly used stock solution

Solution	Mathod of preparation
M Adenosine Triphosphate (ATP)	60.0 mg of ATP was dissolved in 0.8 ml of distilled water. The pH was adjusted to 7.0 with 0.1 N NaOH and volume made upto 1 ml with distilled water. The solution was dispensed into small aliquots and stored at – 70°C.
1 M CaCl	54.0 g $CaCl_2.2H_2O$ was dissolved in 200 ml of pure water (Mill-Q or equivalent). The solution was sterilized by passing through a 0.22 micron filter and stored in 1 ml aliquots at 4°C.
0.5 M EDTA (pH 8.0)	186.1 g of ethylenediamine tetra acetic acid disodium salt, $2H_2O$ was added to 800 ml of distilled water, stirred vigorously on a magnetic stirrer, pH was adjusted to 8.0 with NaOH (20.0 g of NaOH pellets). Volume made upto 1 L with distilled water, dispensed into aliquots and sterilized by autoclaving.
Ethidium bromide (10 mg ml^{-1})	1.0 g of ethidium bromide was added to 100 ml of distilled water and stirred on a magnetic stirrer for several hours to ensure that the dye has dissolved. The solution was transferred to a dark bottle and stored at room temperature.
Phenol: Chloroform: Isoamyl alcohol	Buffer saturated phenol, chloroform and isoamyl alcohol were mixed in the ratio of 25:24:1. The equilibrated mixture was stored under a layer of 0.01 M Tris-HCl (pH 7.6) at 4°C dark glass bottle
Chloroform: Isoamyl alcohol	Buffer saturated chloroform and isoamyl alcohol were mixed in the ratio of 24:1.
IPTG (Isopropyl- β-D-thiogalactopyranoside)	A solution of IPTG was made by dissolving 2.0 g of IPTG in 8 ml of distilled water. Volume was made upto 10 ml with distilled water and sterilized by filtration through a 0.22 micron disposable filter. The solution was dispensed into 1 ml of aliquots and stored at -20°C.

Contd.

Solution	Mathod of preparation
1 M $MgCl_2$	203.3 g of $MgCl_2.6H_2O$ was dissolved in 800 ml distilled water. The volume was made upto 1 L, dispensed into aliquots and sterilized by autoclaving.
3 M Solution acetate (pH 4.8)	408.1 g of $NaOAc.3H_2O$ was dissolved in 800 ml distilled water. The pH was adjusted to 4.8 with glacial acid. Volume made upto 1 L with distilled water, dispensed into aliquots and sterilized by autoclaving.
5 M NaCl	233.8 g of NaCl was dissolved in 800 ml of distilled water, volume made upto 1 L with distilled water, dispensed into aliquots and sterilized by autoclaving.
10% Solution dodecyl sulphate (SDS)	100 g of electophoresis grade SDS was dissolved in 900 ml distilled water, heated at 68°C to assist dissolution and pH was adjusted to 7.2 by adding few drops of concentrated HCl. The volume was made upto 1 L with distilled water, dispensed in aliquots.
10 (N) NaOH	Dissolve 400 g of NaOH in 800 ml of distilled water and make up the volume to 1 L with distilled water.
1 M tris HCL	121.1 g of Tris. Base was dissolved in 800 ml of distilled water. pH was adjusted to the desired value by adding concentrated HCl (for pH 7.4, HCl 70 ml; for pH 8.0, HCl 42 ml). The solution was allowed to cool down to room temperature before making final adjustment to the pH. The volume was made up to 1 litre with distilled water, dispensed into aliquots and sterilized by autoclaving.
Chloro-3-indolyl-β-D-galactopyranoside	Dissolving X-gal in dimethyl formamide to make 20 mg /ml solution and stored at -20°C.

(ix) Materials, source, catalogue mumber of chemicals and reagents

Material	Source	Catalogue No.
Vector		
+pGEMT-easy	Promega	
+pBI 121	Clonetech	1996 /1997
Cells		
+*Escherichia coli* DH 5α	Stratagene, USA	200233
+*Agrobacterium. tumefaciens* LBA 4404	(Hoekema *et al.*, 1983)	
Marker		
+ 1kb DNA ladder	MBI Fermentas	SM 0311
Enzymes		
+Restroctopn endonucleases	MBI Fermentas	GA 2151
+RNase A	Sigma chemical Co.	R5000
+Proteinase K	Ambion	2548
+T4 Ligase	Amersham,USA	E2050Y
Radioisotope (α-^{32}P) dCTP sp. Activity 4000 Ci m/Mol	Board of Radiation and Isotope Technology (BRIT), Bombay	LCP 102
Nitrocellulose Membrane (NCM)	Amersham, USA	088120
5-bromo-4-chloro-3-indolyl-β-D- galacto pyranoside (X gal)	USB	12385
Isopropylthio-β-D-galactoside (IPTG)	USB	17886
Antibiotics		
+Ampicillin (Sodium)	Sigma Chemical Co.	A0104-10 g
+Kanamycin monosulf.		K0126-5 g
+Rifampicin		R0146-1 g
+Cefotaxime sodium		C0111-5 g

Contd.

Material	Source	Catalogue No.
Kits		
+Random primer labeling kit	Bangalore genei Pvt. Ltd.	KT-04
+RNeasy plant mini kit	QIAGEN, USA	74104
+Gel extraction kit	QIAGEN, USA	28704
+Taq PCR core kit (250)	QIAGEN, USA	201223
+Omniscript RT kit (50)	QIAGEN, USA	205111
+DNeasy plant mini kit	QIAGEN, USA	69104

APPENDIX III

Stock solutions for Murashige and Skoog's medium (MS)

Constitutions	Amount ($mg\ l^{-1}$)
Stock solution I	
-NH_4NO_3	33000
-KNO_3	38000
-$CaCl_2.2H_2O$	8800
-$MgSO_4.7H_2O$	7400
-KH_2PO_4	3400
Stock solution II	
-KI	166
-H_3BO_3	1240
-$MnSO_4.7H_2O$	4460
-$ZnSO_4.7H_2O$	1720
-$Na_2MoO_4.2H_2O$	50
-$CuSO_4.5H_2O$	5
-$CoCl_2.6H_2O$	5
Stock solution III	
-$FeSO_4.7H_2O$	5560
-$Na_2.EDTA.2H_2O$	7460
Stock solution IV	
-Inositol	20000
-Nicotinic acid	100
-Pyridoxine HCL	100
-Thiamine HCL	100
-Glycine	400

To prepare 1 lit. of medium take 50 ml of stock 1, 5 ml of stock II, 5 ml of stock III, and 5 ml of stock IV.

Dissolve $FeSO_4.7H_2O$ and $Na_2.EDTA.2H_2O$ separately in 450 ml distilled water by heating and constant stirring. Mix the two solutions, adjust the pH to 5.7 and add distilled water to make up the final volume to 1 litre.

To make MT medium, make MS medium first with the above solution and then add myo-inositol (100 mg/l), nicotinic acid (vitamin B 3) (5 mg/l), pyridoxine HCl (Vit. B6) (10 mg/l), thiamin HCl (Vit. B1) (10 mg/l), glycine (2 mg/l), Sucrose (5%).

Antibiotic solutions for construction of CP gene and transformation

Name	Stock solution (mg/ml)	Solvent
Ampicillin	50	Water
Kanamycin	100	Water
Cefotaxime	250	Water
Rifampicin	50	Methanol

All antibiotics which are water soluble were sterilized by filtration (0.22 µm pore size) and then stored as frozen stock solutions at – 20°C. Antibiotics were added directly to liquid medium or to molten and precooled (45-55°C) agar solidified media.

Modified MS medium for French bean transformation and regeneration

Seed germination medium

MS medium without any hormone

Co-cultivation medium:

MS+1 µg/ml each of 2,4-D & NAA

Selection medium:

(100µg/ml Kan +300 µg/ml /Cef.)

+*Callus formation:*

MS+ 1 µg/ml BA+1 µg/ml NAA

+ *Shooting medium:*

MX + 1 µg/ml BA + 1 µg/ml kinetin

+*Rooting medium:*

MS + 1µg/ml IAA

Modified MT medium for citrus transformation and regeneration.

Seed germination medium

MS medium without any hormone

Co-cultivation medium:

MT + 1μg/ml each of IAA + KT

Selection medium:

MT plus (200 μg/ml Kan + 300 μg/ ml Cef.) and supplement with Hormone:

Culture	Kinnow (mg/l)	Mosambi
+Callus formation:	BA (1) + KT (1)	BA (1) + KT (1)
+Shooting medium:	IAA (1) + KT (2)	IAA (1) + KT (2)
+Rooting medium:	BA (2) + GA (2)	NAA (1)

BA: 6 – benzyladenine; KT: Kinetin; IAA: Indole-3-Acetic Acid; NAA: Naphthalene Acetic Acid; 2,4-D: 2,4-Dichlorophenoxy Acetic Acid.

APPENDIX IV

PREPARATION OF BUFFERS

Notes. Buffers of constant ionic strength for use in electrophoresis have been described by Miller and Golder, *Arch. B*. 29, 420 (1950) and by Green, *J.A.C.S*. 55, 2331 (1933).

It should be noted that with many buffers the pH changes appreciably with temperature. In buffers employing Na^+ or K^+ these may be interchanged with negligible alteration to the pH values.

1. *Glycine-HCl buffer (0.05 M),pH 2.2-3.6*

(Gomori, *Methods in Enzymology,* 1, Academic Press Inc., New York (1955), p. 139; after Sorenson)

Glycine, M. wt. = 75.07; 0.2 M-solution contains 15.01 g in 1,000 ml

pH	0.2 M-*glycine* (ml)	0.2 N-HCl (ml)	
2.2	50	44.0	Dilute to 200 ml with H_2O
2.4	50	32.4	" "
2.6	50	24.2	" "
2.8	50	16.8	" "
3.0	50	11.4	" "
3.2	50	8.2	" "
3.4	50	6.4	" "
3.6	50	5.0	" "

2. *Phthalate-HCl buffer (0.05_M), pH 2.2-3.8*

[Bower and Bates, *J. Res. Nat. Bur. Stand.*, 55, 197 (1955)]

KH phthalate, M. wt. = 204.23; 0.2 M-solution contains 40.85 g in 1,000 ml

pH at 25°C	0.2 M-KH phthalate (ml)	0.2 N-HCl (ml)	
2.2	5	4.95	Dilute to 20 ml with H_2O
2.4	5	4.22	" "
2.6	5	3.54	" "
2.8	5	2.89	" "
3.0	5	2.23	" "
3.2	5	1.57	" "
3.4	5	1.04	" "
3.6	5	0.63	" "
3.8	5	0.29	" "

3. Na_2HPO_4-citric acid buffer, pH 2.2-8.0

[McIlvaine, J.B.C. 49, 183 (1921)]

Cannot be used in the presence of Ca^{++} or Mg^{++}

$Na_2HPO_4.2H_2O$, M. wt. = 178.05; 0.2 M-soln. contains 35.61 g in 1,000 ml.
Citric acid. H_2O. M. wt. 210.14; 0.1 M-soln. contains 21.01 g in 1,000 ml

pH	0.2 M-Na_2HPO_4 (ml)	0.1 M-Citric acid (ml)	pH	0.2 M-Na_2HPO_4 (ml)	0.1 M-Citric acid (ml)
2.2	0.40	19.60	5.2	10.72	9.20
2.4	1.24	18.76	5.4	11.15	8.85
2.6	2.18	17.82	5.6	11.60	8.40
2.8	3.17	16.83	5.8	12.09	7.91
3.0	4.11	15.89	6.0	12.63	7.37
3.2	4.94	15.06	6.2	13.22	6.78
3.4	5.70	14.30	6.4	13.85	6.15
3.6	6.44	13.56	6.6	14.55	5.45
3.8	7.10	12.90	6.8	15.45	4.55
4.0	7.71	12.29	7.0	16.47	3.53
4.2	8.28	11.72	7.2	17.39	2.61
4.4	8.82	11.18	7.4	18.17	1.83
4.6	9.35	10.65	7.6	18.73	1.27
4.8	9.86	10.14	7.8	19.15	0.85
5.0	10.30	9.70	8.0	19.45	0.55

4. *Citric acid-sodium citrate buffer (0.1 M), pH 3.0-6.2*

(N. Hemington and R.M.C. Dawson)

Citric acid H_2O, M. wt. = 210.14; 0.1 M-solution contains 21.01 g in 1,000 ml.Na_3

Citrate $2H_2O$, M. wt. = 294.12; 0.1 M-solution contains 29.4 g in 1,000 ml

pH	0.1M-Citric acid (ml)	0.1M-Na_3citrate (ml)	pH	0.1M-Citric acid (ml)	0.1M-Na_3citrate (ml)
3.0	16.4	3.6	4.8	8.0	12.0
3.2	15.5	4.5	5.0	7.0	13.0
3.4	14.6	5.4	5.2	6.1	13.9
3.6	13.7	6.3	5.4	5.1	14.9
3.8	12.7	7.3	5.6	4.2	15.8
4.0	11.8	8.2	5.8	3.2	16.8
4.2	10.8	9.2	6.0	2.3	17.7
4.4	9.9	10.1	6.2	1.6	18.4
4.6	8.9	11.1			

5. *β β'-Dimethylglutaric acid-NaOH buffer (0.05_M) pH 3.2-7.6*

(Stafford *et al., B.B.A.* 18, 318 (1955); the figures were kindly supplied by R.Hems)

β β'-Dimethylglutaric acid, M. wt. = 160.2; 0.1 M-solution contains 16.02 g in 1,000 ml

pH at 21°	0.1 M-β β'-Dimethylglutaric acid (ml)	0.2 N-NaOH (ml)	
3.2	100	8.3	Dilute to 200 ml with H_2O
3.4	100	14.7	" "
3.6	100	22.0	" "
3.8	100	27.4	" "
4.0	100	33.3	" "
4.2	100	36.8	" "
4.4	100	39.8	" "
4.6	100	41.7	" "
4.8	100	43.9	" "
5.0	100	46.2	" "
5.2	100	49.0	" "

Contd.

pH at 21°	0.1 M-β β'-Dimethylglutaric acid (ml)	0.2 N-NaOH (ml)	
5.4	100	52.0	" "
5.6	100	55.8	" "
5.8	100	59.7	" "
6.0	100	65.0	" "
6.2	100	70.5	" "
6.4	100	75.5	" "
6.6	100	84.7	" "
6.8	100	88.0	" "
7.0	100	90.4	" "
7.2	100	92.1	" "
7.4	100	93.2	" "
7.6	100	94.0	" "

6. *Sodium acetate-acetic acid buffer (0.2 M), pH 3.6-5.8*

[Walpole, J.C.S. 105, 2501 (1914)]

Na acetate, $3H_2O$, M. wt. = 136.09; 0.2 M-soln. contains 27.22 g in 1,000 ml

pH at18° C	0.2 M-Sodium acetate (ml)	0.2 M-Acetic acid (ml)
3.6	0.75	9.25
3.8	1.20	8.80
4.0	1.80	8.20
4.2	2.65	7.35
4.4	3.70	6.30
4.6	4.90	5.10
4.8	5.90	4.10
5.0	7.00	3.00
5.2	7.90	2.10
5.4	8.60	1.40
5.6	9.10	0.90
5.8	9.40	0.60

7. *Succinic acid-NaOH buffer (0.05M), pH 3.8-6.0*

(Gomori, *Methods in Enzymology* ,vol. 1, Academic Press Inc., New York (1955), p. 141)

Succinic acid, M. wt. = 118.09; 0.2 M-solution contains 23.62 g in 1,000 ml

pH	0.2 M-Succinic acid (ml)	0.2 N-NaOH (ml)	
3.8	25	7.5	Dilute to 100 ml with H_2O
			" "
4.0	25	10.0	" "
4.2	25	13.3	" "
4.4	25	16.7	" "
4.6	25	20.0	" "
4.8	25	23.5	" "
5.0	25	26.7	" "
5.2	25	30.3	" "
5.4	25	34.2	" "
5.6	25	37.5	" "
5.8	25	40.7	" "
6.0	25	43.5	" "

8. *Phthalate-NaOH buffer (0.05M), pH 4.1-5.9*

[Bower and Bates, *J. Res. Nat. Bur. Stand.* 55, 197 (1955)]

KH phthalate, M. wt. = 204.23; 0.2 M-solution contains 40.85 g in 1,000 ml

pH at 25°	0.2M-KH phthalate (ml)	0.2N-NaOH (ml)	
4.1	5	0.13	Dilute to 20 ml with H_2O
4.3	5	0.47	" "
4.5	5	0.87	" "
4.7	5	1.36	" "
4.9	5	1.94	" "
5.1	5	2.55	" "
5.3	5	3.16	" "
5.5	5	3.66	" "
5.7	5	4.06	" "
5.9	5	4.37	" "

9. *Sodium cacodylate-HCl buffer* (0.05M), *pH* 5.0-7.4

[Plumel, *Bull. Soc. Chim. biol., Paris* 30, 129 (1948)]

Na$(CH_3)_2$.AsO_2.$3H_2O$, M. wt. = 214.02; 0.1 M-solution contains 21.40 g in 1,000 ml

pH at 15°	0.1 M-Na$(CH_3)_2$. as O_2.$3H_2O$ (ml)	0.1 N-HCl (ml)	
5.0	100	93.5	Dilute to 200 ml with H_2O
5.2	100	90.1	" "
5.4	100	85.2	" "
5.6	100	78.4	" "
5.8	100	69.6	" "
6.0	100	59.1	" "
6.2	100	47.7	" "
6.4	100	36.5	" "
6.6	100	26.6	" "
6.8	100	18.6	" "
7.0	100	12.6	" "
7.2	100	8.3	" "
7.4	100	5.4	" "

10. *Sodium hydrogen maleate—NaOH buffer (0.05M), pH 5.2-6.8*

[Temple, J.A.C.S. 51, 1754 (1929)]

0.2M-NaH maleate is prepared by dissolving in H_2O 23.2 g maleic acid (or 19.6 g maleic anhydride) and 8 g NaOH and making up the volume to 1,000 ml

pH at 25°	0.2M-NaOH (ml)	0.2M-NaH maleate (ml)	
5.2	7.2	50	Dilute to 200 ml with H_2O
5.4	10.5	50	" "
5.6	15.3	50	" "
5.8	20.8	50	" "
6.0	26.9	50	" "
6.2	33.0	50	" "
6.4	38.0	50	" "
6.6	41.6	50	" "
6.8	44.4	50	" "

11. *Na_2HPO_4-NaH_2PO_4 buffer* (0.1M), *pH* 5.8-8.0

Gomori, *Methods in Enzymology,* vol. 1, Academic Press Inc., New York (1955), (p.143; after Sorenson)

$Na_2HPO_4.2H_2O$, M. wt. = 178.05; 0.20 M-solution contains 35.61 g in 1,000 ml
$Na_2HPO_4.12H_2O$, M. wt. = 358.22; 0.2 M-solution contains 71.64 g in 1,000 ml
$NaH_2PO_4.H_2O$, M. wt. = 138.0;0.2 M-solution contains 27.6 g in 1,000 ml
$NaH_2PO_4.2H_2O$, M. wt. = 156.03; M-solution contains 31.21 g in 1,000 ml

pH	0.2M-Na_2HPO_4 (ml)	0.2M-NaH_2PO_4 (ml)	
5.8	8.0	92.0	Dilute to 200 ml with H_2O
6.0	12.3	87.7	" "
6.2	18.5	81.5	" "
6.4	26.5	73.5	" "
6.6	37.5	62.5	" "
6.8	49.0	51.0	" "
7.0	61.0	39.0	" "
7.2	72.0	28.0	" "
7.4	81.0	19.0	" "
7.6	87.0	13.0	" "
7.8	91.5	8.5	" "
8.0	94.7	5.3	" "

12. *KH_2PO_4-NaOH buffer (0.05M), pH 5.8-8.0*

[Bower and Bates, *J. Res. Nat. Bur. Stand.* 55, 1970 (1955)]

For sodium free buffer use KOH.

KH_2PO_4, M. wt. = 136.09; 0.2 M-soln. contains 27.22 g in 1,000 ml

pH at 25°C	0.2M-KH_2PO_4 (ml)	0.2N-NaOH (ml)	
5.8	5	0.36	Dilute to 20 ml with H_2O
6.0	5	0.56	" "
6.2	5	0.81	" "
6.4	5	1.16	" "
6.6	5	1.64	" "
6.8	5	2.24	" "
7.0	5	2.91	" "
7.2	5	3.47	" "
7.4	5	3.91	" "
7.6	5	4.24	" "
7.8	5	4.45	" "
8.0	5	4.61	" "

13. *Sodium arsenate-HCl buffer (0.05M), pH* 5.8-7.6

(J. Lowenstein, Unpublished)

$Na_2HAsO_4.7H_2O$, M. wt. = 312.03; 0.1 M-solution contains 31.2 g in 1,000 ml

$Ha_2HAsO_4.12H_2O$, M. wt. = 402.10; 0.1 M-solution contains 40.21 g in 1,000 ml

pH at 25°C	0.1M-Na_2HAsO_4 (ml)	0.1N-HCl (ml)	
5.8	25	21.6	Dilute to 50 ml with H_2O
6.0	25	20.5	" "
6.2	25	18.9	" "
6.4	25	16.5	" "
6.6	25	12.7	" "
6.8	25	9.0	" "
7.0	25	6.2	" "
7.2	25	4.1	" "
7.4	25	2.6	" "
7.6	25	1.6	" "

14. *Sodium bicarbonate-5 % CO_2 buffer, pH 6.0-8.0*

Calculated from the Henderson-Hasselbach equation

$$pH = pK' + \log$$

Concentrations of bicarbonate are approximately correct when atmospheric pressure varies between 725 and 760 mm Hg.

Temperature. For every degree decrease in temperature down to 20° the bicarbonate conc. is increased by approximately 1.88 % to give the same pH as at 37°C.

pH at 37°C	Conc. of $NaHCO_3$ (M. wt. = 84.02) in equilibrium with gas phase containing 5% CO_2
6.0	5.86×10^{-4}M
6.2	9.29×10^{-4}M
6.4	1.47×10^{-3}M
6.6	2.33×10^{-3}M
6.8	3.70×10^{-3}M
7.0	5.86×10^{-3}M
7.2	9.29×10^{-3}M

Contd.

pH at 37°C	Conc. of $NaHCO_3$ (M. wt. = 84.02) in equilibrium with gas phase containing 5% CO_2
7.4	1.47×10^{-2}M
7.6	2.33×10^{-2}M
7.8	3.70×10^{-2}M
8.0	5.86×10^{-2}M

15. *Imidazole (glyoxaline)-HCl buffer (0.05M), pH* 6.2-7.8

[Mertz and Owen, *P.S.E.B.M.* 43, 204 (1940)]

Imidazole , M. wt = 68.08; 0.2 M-solution contains 13.62 g in 1,000 ml

pH at 25°C	0.2M-imidazole (ml)	0.1N-HCl (ml)	
6.2	25	42.9	Dilute to 100 ml with H_2O
6.4	25	39.8	" "
6.6	25	35.5	" "
6.8	25	30.4	" "
7.0	25	24.3	" "
7.2	25	18.6	" "
7.4	25	13.6	" "
7.6	25	9.3	" "
7.8	25	6.0	" "

16. 2:4:6- *Collidine –HCl buffer (0.05 M), pH* 6.4-8.3

[Gomori, *P.S.E.B.M.* 62, 33 (1946)]

2:4:6-Collidine, M. wt. = 121.2; density, 0.917; 0.2 M-solution contains 24.24 g in 1,000 ml

pH		0.2M-2:4:6-*Collidine* (ml)	0.1N HCl (ml)	
23°C	37°C			
6.45	6.37	25	45.0	Dilute to 100 ml with H_2O
6.62	6.54	25	42.5	" "
6.80	6.72	25	40.0	" "
6.92	6.84	25	37.5	" "
7.03	6.95	25	35.0	" "

Contd.

pH		**0.2M-2:4:6-*Collidine* (ml)**	**0.1N HCl (ml)**	
23°C	**37°C**			
7.13	7.05	25	32.5	" "
7.22	7.14	25	30.0	" "
7.30	7.23	25	27.5	" "
7.40	7.32	25	25.0	" "
7.50	7.40	25	22.5	" "
7.57	7.50	25	20.0	" "
7.67	7.60	25	17.5	" "
7.77	7.70	25	15.0	" "
7.88	7.80	25	12.5	" "
8.00	7.94	25	10.0	" "
8.18	8.10	25	7.5	" "
8.35	8.28	25	5.0	" "

17. *Sodium diethylbarbiturate (veronal)-HCl buffer pH* 6.8-9.6

(Britton and Robinson, J.C.S. 1931, 1456)

Na diethylbarbiturate, M. wt. = 206.2; 0.04 M-solution contains 8.25 g in 1,000 ml

pH at 18°C	**0.04M-sodium Diethylbarbiturate (ml)**	**0.2N-HCl (ml)**
6.8	100	18.4
7.0	100	17.8
7.2	100	16.7
7.4	100	15.3
7.6	100	13.4
7.8	100	11.47
8.0	100	9.39
8.2	100	7.21
8.4	100	5.21
8.6	100	3.82
8.8	100	2.52
9.0	100	1.65
9.2	100	1.13
9.4	100	0.70
9.6	100	0.35

18. *N-ethylmorpholine-HCl buffer (0.1M), pH 7.0-8.2*

(Data kindly supplied by Prof. A. Albert)

N-Ethylmorpholine; M. wt. = 115.17; 0.2 M-solution contains 23.03 g in 1,000 ml. *N*-Ethylmorpholine should be ddried and freshly distilled (B.p. 138-9°, 763 mm)

pH at 20°C	0.2M-*N*-Ethylmorphine	N-HCl	
7.0	50	8.0	Dilute to 100 ml with H_2O
7.2	50	7.1	" "
7.4	50	6.1	" "
7.6	50	5.0	" "
7.8	50	4.0	" "
8.0	50	2.9	" "
8.2	50	2.0	" "

19. *Triethanolamine-HCl buffer (0.005M), pH 7.1-8.6*

(R. Hems. Unpublished)

Triethanolamine, M. wt. = 149.2; 0.01 M-solution contains 1.49 g in 1,000 ml. For purification of triethanolamine *see J.A.C.S.* 55, 4150 (1933).

Where NaCl is not objected to in the buffer it is more convenient to use triethanol-amine HCl-NaOH buffers (*see* next Table).

pH at 25°C	0.01M-Triethanolamine (ml)	0.01N-HCl (ml)	
7.10	10	8.13	Dilute to 20 ml with H_2O
7.35	10	7.11	" "
7.57	10	6.10	" "
7.76 (7.56 at 40°)	10	5.08	" "
7.94	10	4.06	" "
8.13	10	3.05	" "
8.37	10	2.03	" "
8.62	10	1.02	" "

20. *Triethanolamine hydrochloride-NaOH buffer (0.05M), pH 6.8-8.6*

(R. Hems, unpublished)

Triethanolamine hydrochloride, M. wt. = 185.7; 0.1 M-solution contains 18.57 g in 1,000 ml.

The hydrochloride is prepared by dissolving triethanolamine in 1 vol. of ethanol, neutralizing with conc. HCl and cooling to 0°. M.p. 178-9°. *See* previous table for NaCl- free buffer.

pH at 20°C	**0.1M-Triethanolamine hydrochloride (ml)**	**0.1N-NaOH (ml)**	
6.8	10	1.20	Dilute to 20 ml with H_2O
7.0	10	1.80	" "
7.2	10	2.43	" "
7.4	10	3.16	" "
7.6	10	4.01	" "
7.8	10	5.20	" "
8.0	10	6.30	" "
8.2	10	7.20	" "
8.4	10	8.00	" "
8.6	10	9.20	" "

21. *Tris-(hydroxymethyl)-aminomethane-HCl (Tris-HCl) Buffer (0.05M), pH 7.2-9.1*

(Sigma Chemical Co. Bulletin, 106)

Tris-(hydroxymethyl)-aminomethane, M. wt. = 121.14; 0.2M-soln. contains 24.23 g in 1,000 ml

pH		**0.2M-*tris* (ml)**	**0.1N-HCl (ml)**	
23°C	**37°C**			
9.10	8.95	25	5	Dilute to 100 ml with H_2O
8.92	8.78	25	7.5	" "
8.74	8.60	25	10.0	" "
8.62	8.48	25	12.5	" "
8.50	8.37	25	15.0	" "
8.40	8.27	25	17.5	" "
8.32	8.18	25	20.0	" "
8.23	8.10	25	22.5	" "
8.14	8.00	25	25.0	" "
8.05	7.90	25	27.5	" "

Contd.

pH		0.2M-*tris* (ml)	0.1N-HCl (ml)	
23°C	37°C			
7.96	7.82	25	30.0	" "
7.87	7.73	25	32.5	" "
7.77	7.63	25	35.0	" "
7.66	7.52	25	37.5	" "
7.54	7.40	25	40.0	" "
7.36	7.22	25	42.5	" "
7.20	7.05	25	45.0	" "

22.Boric acid-borax buffer (0.2M in terms of borate), pH 7.4-9.0

[Holmes, *Anat. Record* 86, 157 (1943)]

Borax, $Na_2B_4O_7$. 10 H_2O, M. wt. = 381.43; 0.05 M-solution (= 0.2M borate) contains 19.07 g in 1,000 ml.

Boric acid, M. wt. = 61.84; 0.2M-soln. contains 12.37 g in 1,000 ml.

Borax, $Na_2B_4O_7$.10H_2O, may lose water of crystallization and it should be kept in a stoppered bottle. Solns. of borax can be prepared also by half-neutralizing solns. of boric acid.

pH	0.05M-borax (ml)	0.2M-boric acid (ml)
7.4	1.0	9.0
7.6	1.5	8.5
7.8	2.0	8.0
8.0	3.0	7.0
8.2	3.5	6.5
8.4	4.5	5.5
8.7	6.0	4.0
9.0	8.0	2.0

23. 2-Amino-2 methyl-1:3-propanediol-HCl buffer (0.05M), pH 7.8-9.7

[Gomori,*P.S.E.B.M.* 62, (1946)]

2-Amino-2-methyl-1:3-propanediol, M. wt. = 105.15; 0.2 M-solution contains 21.03 g in 1,000 ml.

pH		0.2M-2-amino-2-methyl-1:3-propanediol (ml)	0.1N-HCl (ml)	
23°C	37°C			
9.72	9.62	25	5.0	Dilute to 100 ml with H_2O
9.56	9.45	25	7.5	" "
9.38	9.27	25	10.0	" "
9.26	9.15	25	12.5	" "
9.15	9.03	25	15.0	" "
9.05	8.94	25	17.5	" "
8.96	8.85	25	20.0	" "
8.87	8.76	25	22.5	" "
8.78	8.67	25	25.0	" "
8.70	8.58	25	27.5	" "
8.60	8.50	25	30.0	" "
8.50	8.40	25	32.5	" "
8.40	8.30	25	35.0	" "
8.30	8.20	25	37.5	" "
8.18	8.07	25	40.0	" "
8.00	7.90	25	42.5	" "
7.83	7.72	25	45.0	" "

24. *Diethanolamine-HCl buffer (0.05M), pH 8.0-10.0*

(J. Lowenstein, unpublished)

Diethanolamine, M. wt. = 105.1; 0.2 M-solution contains 21.02 g in 1,000 ml

B.p. $142\text{-}3^{1\text{mm Hg}}$

pH at 25°C	0.2M-diethanolamine (ml)	0.1N-HCl (ml)
10.0	25	3.6
9.7	25	6.9
9.5	25	11.1
9.3	25	15.6
9.1	25	20.4
8.9	25	27.1
8.7	25	32.7
8.5	25	37.7
8.3	25	42.0
8.0	25	45.9

25. *Glycine-NaOH buffer (0.05M), pH8.6-10.6*

(Gomori, *Methods in Enzymology,* vol. 1, Academic Press Inc. New York (1955), p. 145; after Sorenson)

Glycine, M. wt. = 75.07; 0.2 M-solution contains 15.01 g in 1,000 ml

pH	0.2M-glycine (ml)	0.2N-NaOH (ml)	
8.6	50	4.0	Dilute to 200 ml with H_2O
8.8	50	6.0	" "
9.0	50	8.8	" "
9.2	50	12.0	" "
9.4	50	16.8	" "
9.6	50	22.4	" "
9.8	50	27.2	" "
10.0	50	32.0	" "
10.4	50	38.6	" "
10.6	50	45.5	" "

26. *Borax-NaOH buffer (0.05M-borate), pH 9.3-10.1*

(Gomori, *Methods in Enzymology,* vol. 1, Academic Press Inc., New York (1955), p. 146; after Clark and Lubs)

See Table 22

pH	0.05M-borax (ml)	0.2N-NaOH (ml)	
9.3	50	0.0	Dilute to 200 ml with H_2O
9.4	50	11.0	" "
9.6	50	23.0	" "
9.8	50	34.0	" "
10.0	50	43.0	" "
10.0	50	46.0	" "

27. *Sodium carbonate-bicarbonate buffer (0.1M), pH 9.2-10.8*

[Deloty and King, *B.J.* 39, 245 (1945)]

Cannot be used in the presence of Ca^{++} or Mg^{++}

$Na_2CO_3.10H_2O$. M. wt. = 286.2; 0.1 M-solution contains 28.62 g in 1,000 ml.

$NaHCO_3$, M. wt = 84.0; 0.1 M-solution contains 8.40 g in 1,000 ml

pH		0.1M-Na_2CO_3 (ml)	0.1M-$NaHCO_3$ (ml)
20°C	**37°C**		
9.16	8.77	1	9
9.40	9.12	2	8
9.51	9.40	3	7
9.78	9.50	4	6
9.90	9.72	5	5
10.14	9.90	6	4
10.28	10.08	7	3
10.53	10.28	8	2
10.83	10.57	9	1

Source: By R.M.C. Dawson (A.R.C. Institute of Animal Physiology, Babram, Combsw.) and W.H. Elliott (Dept of Biochemistry, University of Oxford)

STANDARD pH SOLUTIONS

The British Standards Institution (Publication B.S. 1647:1950) recommend the following solutions for calibration of glass electrodes:

Primary Standard

0.05M-potassium hydrogen phthalate (10.21 g of pure, dry KH phthalate in freshly distilled water, or recently boiled and cooled distilled water, and made up to 1,000 ml); the pH of this solution *is difined* as 4.000 at 15°, 4.005 at 25°, and 4.026 at 38°C.

Secondary Standards

1. 0.1M-potassium tetroxalate? [$KH_3(C_2O_4)_2.2H_2O$, M. wt. = 245.19].pH = 1.48 at 25° and 1.50 at 38°C.
2. 0.01N-HCl + 0.09M-KCl. pH = 2.07 at 25° and 2.08 at 38°C.
3. 0.1M-CH_3COOH and 0.1M-CH_3COONa (prepared from acetic acid half neutralized with NaOH, not from sodium acetate). pH = 4.65 at 12°, 4.64 at 25° and 4.65 at 38°C.
4. 0.025M-KH_2PO_4 + 0.025M-$Na2HPO_4.2H_2O$. pH = 6.85 at 25° and 6.84 at 38°C.
5. 0.05M-$Na_2B_4O_7.10H_2O$‡ (M.wt. = 381.43). pH = 9.18 at 25° and 9.07 at 38°C.

The National Bureau of Standards (U.S.A.) defines the pH scale in terms of the following solutions [data from Bates, *Electrometric pH Determinations*, Wiley & Sons Inc., New York (1954), p. 118].

1. 0.05M-potassium tetroxalate† [12.71 g $KH_3(C_2O_4)_2.2H_2O$per l.]. pH = 1.68 at 20-25° and 1.69 at 30°C.
2. A solution of potassium hydrogen tartrate (saturated at 25°C). pH = 3.56 at 25° and 3.55 at 30°C.
3. 0.05M-potassium hydrogen phthalate (10.211 g per l). pH = 4.00 at 20° and 4.01 at 25-30°C.

4. 0.025M-KH_2PO_4 + 0.025M-Na_2HPO_4 (3.402 g KH_2PO_4 + 4.451 g $Na_2HPO_4.2H_2O$ per l). pH = 6.88 at 20°, 6.86 at 25°, and 6.85 at 30°C.

5. 0.01M-borax (3.814 g $Na_2B_4O_7.10H_2O$‡ per l). pH = 9.22 at 20°, 9.18 at 25°, and 9.14 at 30°C.

† Dry below 60°C.

APPROXIMATE pK_a VALUES OF SUB-STANCES USEFUL FOR BUFFER SYSTEMS

The approximate composition of buffers can be calculated from the equation

$$\mathrm{pH} = \mathrm{p}K + \log \frac{\text{Conc. of salt}}{\text{Conc. of acid}}$$

Where accuracy is required the pH of such solution should be determined before use.

A buffer table is given in the following section for those compounds marked with an asterisk.

pK_a	Compound	pK_a	Compound
1.3	k_1 Oxalic acid	6.4	k_3 Citric acid*
1.92	k_1 Maleic acid	6.4	k_3 4- or 5-Hydroxy-
1.96	k_1 Phosphoric acid		methyl imidazole
2.45	k_1 Glycine*	6.54	k_2 Pyrophosphoric acid
2.9	k_1 Phthalic acid	6.6	Arsenic acid*
3.02	k_1 Fumaric acid	6.7	Phosphoric acid*
3.03	k_1 Tartaric acid	6.95	Imidazole*
3.1	k_1 Citric acid	$7.32^{37°}$	2:4:6-Collidine*
3.15	k_1 Gly.....Glycine	7.52	4- or 5-Methyl imidazole
3.17	k_1 Furoic acid	7.65	*N*-Ethyl morpholine*
3.66	k_1 β β'-Dimethylglutaric	7.8	Triethanolamine*
	acid*	7.98	Diethylbarbituric acid*
3.75	k_1 Formic acid	8.08	Tris-(hydroxymethyl)-
3.98	k_1 Barbituric acic		aminomethane*
4.18	k_1 Succinic acid*	8.13	k_2 Glycylglycine
4.2	k_1 Benzoic acid	8.36	2:4- or 2:5-Dimethyl
4.26	k_2 Oxalic acid		imidazole

Contd.

pK_a	Compound	pK_a	Compound
4.37	k_2 Tartaric acid	8.44	k_4 Pyrophosphoric acid
4.4	k_2 Fumaric acid	$8.67^{37°}$	2-Amino-2-methyl-1:3-propanediol*
4.73	k_2 Acetic acid*		
4.75	k_2 Citric acid*	8.9	Diethanolamine*
5.05	k_2 Malic acid	9.23	Boric acid*
5.4	k_2 Phthalic acid*	9.25	Ammonium ion
5.6	k_2 Succinic acid*	9.5	Ethanolamine
6.1	k_2 Histidine	9.6	k_2 Glycine*
6.15	k_2 Cacodylic acid*	9.87	Trimethylamine
6.2	k_2 β β'-Dimethy-lglutaric acid*	10.7	Methylamine
6.22	k_2 Maleic acid*		
6.35	k_2 Carbonic acid*		

CONVERSION TABLES

TEMPERATURE

To convert a temperature, in either Centigrade (Celsius) or Fahrenheit, to the other scale, find that temperature in the center column, and then find the equivalent temperature in the other scale either in the Centigrade column to the left or in the Fahrenheit column to the right. For example, if the Centigrade temperature (in the righthand column) is 111.2°. If the *Fahrenheit* temperature is 44°, the equivalent Centigrade temperature (in the lefthand column) is 6.67°.

On the Centigrade scale the temperature of melting ice is 0° and that of boiling water is 100° at normal atmospheric pressure. On the Fahrenheit scale, the equivalent temperatures are 32° and 212° respectively. The formula for converting Centigrade to Fahrenheit is C = 5/9 F-32, and the formula for converting Fahrenheit to Centigrade is F = 9/5 C + 32.

C	C or F	F	C	C or F	F	C	C or F	F	C	C or F	F
-73.33	-100	-148.0	-6.67	20	68.0	15.6	60	140.0	43	110	230
-70.56	-95	-139.0	-6.11	21	69.8	16.1	61	141.8	49	120	248
-67.78	-90	-130.0	-5.56	22	71.6	16.7	62	143.6	54	130	266
-65.00	-85	-121.0	-5.00	23	73.4	17.2	63	145.4	60	140	284
-62.22	-80	-112.0	-4.44	24	75.2	17.8	64	147.2	66	150	302
-59.45	-75	-103.0	-3.89	25	77.0	18.3	65	149.0	71	160	320
-56.67	-70	-94.0	-3.33	26	78.8	18.9	66	150.8	77	170	338
-53.89	-65	-85.0	-2.78	27	80.6	19.4	67	152.6	82	180	356
-51.11	-60	-76.0	-2.22	28	82.4	20.0	68	154.4	88	190	374
-48.34	-55	-67.0	-1.67	29	84.2	20.6	69	156.2	93	200	392
-45.56	-50	-58.0	-1.11	30	86.0	21.1	70	158.0	99	210	410
-42.78	-45	-49.0	-0.56	31	87.8	21.7	71	159.8	100	212	414
-40.0	-40	-40.0	-0	32	89.6	22.2	72	161.6	104	220	428
-37.23	-35	-31.0				22.8	73	163.4	110	230	446
-34.44	-30	-22.0	0.56	33	91.4	23.3	74	165.2	116	240	464
-31.67	-25	-13.0	1.11	34	93.2	23.9	75	167.0	121	250	482

Contd.

C	C or F	F	C	C or F	F	C	C or F	F	C	C or F	F
-28.89	-20	-4.0	1.67	35	95.0	24.4	76	168.8	127	260	500
-26.12	-15	5.0	2.22	36	96.8	25.0	77	170.6	132	270	518
-23.33	-10	14.0	2.78	37	98.6	25.6	78	172.4	138	280	536
-20.56	- 5	23.0	3.33	38	100.4	26.1	79	174.2	143	290	554
-17.8	0	32.0	3.89	39	102.2	26.7	80	176.0	149	300	572
			4.44	40	104.0	27.2	81	177.8	154	310	590
-17.2	1	33.8	5.00	41	105.8	27.8	82	179.6	160	320	608
-16.7	2	35.6	5.56	42	107.6	28.3	83	181.4	166	330	626
-16.1	3	37.4	6.11	43	109.4	28.9	84	183.2	171	340	644
-15.6	4	39.2	6.67	44	111.2	29.4	85	185.0	177	350	662
-15.0	5	41.0	7.22	45	113.0	30.0	86	186.8	182	360	680
-14.4	6	42.8	7.78	46	114.8	30.6	87	188.6	188	370	698
-13.9	7	44.6	8.33	47	116.6	31.1	88	190.4	193	380	716
-13.3	8	46.4	8.89	48	118.4	31.7	89	192.2	199	390	734
-12.8	9	48.2	9.44	49	120.2	32.2	90	194.0	204	400	752
-12.2	10	50.0	10.0	50	122.0	32.8	91	195.8	210	410	770
-11.7	11	51.8	10.6	51	123.8	33.3	92	197.6	216	420	788
-11.1	12	53.6	11.1	52	125.6	33.9	93	199.4	221	430	806
-10.6	13	55.4	11.7	53	127.4	34.4	94	201.2	227	440	824
-10.0	14	57.2	12.2	54	129.2	35.0	95	203.0	232	450	842
-9.44	15	59.0	12.8	55	131.0	35.6	96	204.8	238	460	860
-8.89	16	60.8	13.3	56	132.8	36.1	97	206.6	243	470	878
-8.33	17	62.6	13.9	57	134.6	36.7	98	208.4	249	480	896
-7.78	18	64.4	14.4	58	136.4	37.2	99	210.2	254	490	914
-7.22	19	66.2	15.0	59	138.2	37.8	100	212.0	260	500	932

AREA

1 square centimeter	=	0.155 sq inch
	=	100 sq millimeters
1 square meter	=	1,550 aq inches
	=	10.764 sq feet
	=	1.196 sq yards
	=	10,000 sq centimeters
1 square kilometer	=	0.3861 sq mile
	=	1,000,000 sq meters
1 hectare	=	2,471 acres
	=	10,000 sq meters

Imperial

1 square inch	=	6.452 sq centimeters
	=	1/144 sq foot
	=	1/1296 sq yard
1 square foot	=	929.088 sq centimeters
	=	0.0929 sq meter
1 square yard	=	8,361.3 sq centimeters
	=	0.8361 sq meter
	=	1,296 sq inches
	=	9 sq feet
1 square mile	=	2.59 sq kilometers
	=	640 acres
1 acre	=	0.4047 hectare
	=	43,560 sq feet
	=	4,840 sq yards
	=	4,046.87 sq meters

LENGTH

Metric

1 millimicron	=	0.001 micron
1 micron	=	0.001 millimeter
1 millimeter	=	0.001 meter
	=	0.0394 inch
1 centimeter	=	10 millimeters
	=	0.3937 inch
1 meter	=	39.37 inches
	=	3.281 feet
	=	1000 milimeters
	=	100 centimeters
	=	1.2 varas
1 kilometer	=	3,281 feet
	=	1,094 yards
	=	0.621 mile
	=	1,000 meter

Imperial

1 inch	=	25.4 millimeters
	=	2.54 centimeters
1 foot	=	30.48 centimeters
	=	0.3048 meter
	=	12 inches

1 yard	=	0.9144 meter
	=	91.44 centimeters
	=	3 feet
1 mile	=	1,609.347 meters
	=	1.609 kilometers
	=	5,280 feet
	=	1,760 yards

WEIGHT

Metric

1 milligram	=	0.001 gram
	=	0.0154 grain
1 centigram	=	0.01 gram
	=	0.1543 grain
1 gram	=	0.0353 avoirdupois ounce
	=	15.4324 grains
1 kilogram	=	1,000 grams
	=	353 avoirdupois pounds
	=	2.2046 avoirdupois pounds
1 metric ton	=	1,000 kilograms
	=	2,204.6 pounds
	=	1.102 short tons
	=	0.984 long ton

Imperial

1 grain	=	1/700 avoirdupois pound
	=	0.064799 gram
1 ounce (avoirdupois)	=	28.3496 grams
	=	0.45369 kilogram
	=	1/16 pound
1 pound (avoirdupois)	=	453.593 grams
	=	0.45369 kilograms
	=	16 ounces
1 short ton	=	907.184 kilograms
	=	0.9072 metric ton
	=	2,000 pounds

YIELD

Metric

1 kilogram per hectare	=	0.89 pound per acre
1 cubic meter per hectare	=	14.2916 cubic feet per acre

Imperial

1 pound per acre	=	1.121 kilograms per hectare
1 ton (2,000 lb) per acre	=	2.242 metric tons per hectare
1 cubic foot per acre	=	0.0699 cubic meter per hectare
1 bushel (60 lb) per acre	=	67.26 kilograms per hectare

VOLUME

Metric

1 litre	=	1.057 U.S. quarts liquid
	=	0.9081 quart, dry
	=	0.2642 U.S. gallon
	=	0.221 Imperial gallon
	=	1,000 milliliters or cc
	=	0.0353 cubic foot
	=	61.02 cubic inches
	=	0.001 cubic meter
1 cubic meter	=	61,023.38 cubic inches
	=	35.314 cubic feet
	=	1.308 cubic yards
	=	264.17 U.S. gallons
	=	1,000 liters
	=	28.38 U.S. bushels
	=	1,000,000 cu. centimeters
	=	1,000,000,000 cu. millimeters

Imperial

1 fluid ounce	=	1/128 gallon
	=	29.57 cubic centimeters
	=	29.562 milliliters
	=	1.805 cubic inches
	=	0.0625 U.S. pint (liquid)
1 U.S. quart liquid	=	946.3 milliliters
	=	57.75 cubic inches
	=	32 fluid ounces
	=	4 cups
	=	¼ gallon
	=	2 U.S. pints (liquid)
	=	0.946 litre
1 quart dry	=	1.1012 litres
	=	67.20 cubic inches
	=	2 pints (dry)
	=	0.125 peck
	=	1/32 bushel

1 cubic inch	=	16.387 cubic centimeters
1 cubic foot	=	28,317 cubic centimeters
	=	0.0283 cubic meter
	=	28.316 litres
	=	7.481 U.S. gallons
	=	1,728 cubic inches
1 U.S. gallon	=	16 cups
	=	3.785 litres
	=	231 cubic inches
	=	4 U.S. quarts liquid
	=	8 U.S. pints liquid
	=	8.3453 pounds of water
	=	128 fluid ounces
	=	0.8327 British Imperial gallon
1 British Imperial gallon	=	4.546 liters
	=	1.201 U.S. gallons
	=	277.42 cubic inches
1 U.S. bushel	=	35.24 litres
	=	2,150.42 cubic inches
	=	1.2444 cubic feet
	=	0.03524 cubic meter
	=	2 pecks
	=	32 quarts (dry)
	=	64 pints (dry)

NUMBER OF TREES PER ACRE AT GIVEN SPACINGS*

	No. of		No. of		No. of
Feet	Plants	Feet	Plants	Feet	Plants
6 × 1	7,260	10 × 1	4,356	18 × 4	605
6 × 2	3,630	10 × 2	2,178	18 × 6	404
6 × 3	2,420	10 × 3	1,452	18 × 8	303
6 × 4	1,815	10 × 4	1,089	18 × 10	242
6 × 5	1,452	10 × 5	871	18 × 12	202
6 × 6	1,210	10 × 6	726	18 × 14	173
		10 × 7	622	18 × 16	152
7 × 1	6,223	10 × 8	544	18 × 18	132
7 × 2	3,111	10 × 9	484		
7 × 3	2,074	10 × 10	435		
7 × 4	1,556			20 × 8	272
7 × 5	1,224	12 × 2	1,815	20 × 10	218
7 × 6	1,037	12 × 4	907	20 × 12	184
7 × 7	889	12 × 6	605	20 × 14	156
		12 × 7	454	20 × 16	136
8 × 1	5,445	12 × 10	363	20 × 18	121
8 × 2	2,722	12 × 12	302	20 × 20	109
8 × 3	1,815				
8 × 4	1,361	14 × 2	1,556		
8 × 5	1,089	14 × 4	778	24 × 12	151
8 × 6	907	14 × 6	518	24 × 16	114
8 × 7	778	14 × 8	389	24 × 20	92
8 × 8	680	14 × 10	311	24 × 24	76
		14 × 12	259		
9 × 1	4,840	14 × 14	222		
9 × 2	2,420			30 × 20	72

Contd.

Feet	No. of Plants	Feet	No. of Plants	Feet	No. of Plants
9 × 3	1,613	16 × 2	1,361	30 × 30	48
9 × 4	1,210	16 × 4	680	30 × 40	36
9 × 5	968	16 × 6	454		
9 × 6	807	16 × 8	340		
9 × 7	691	16 × 10	272	40 × 40	27
9 × 8	605	16 × 12	194		
9 × 9	528	16 × 14	170		
		16 × 16			

*In order to obtain the number of plants per acre, divide 43.560 by the product of the spacing in the rows expressed in feet

NUMBER OF TREES PER HECTARE AT GIVEN SPACINGS*

No. of		No. of		No. of	
Meters	Plants	Meters	Plants	Meters	Plants
2 × 0.25	20,000	5 × 1.0	2,000	8 × 1	1,250
2 × 0.50	10,000	5 × 1.5	1,333	8 × 2	555
2 × 0.75	6,000	5 × 2.0	1,000	8 × 3	416
2 × 1.0	5,000	5 × 2.5	800	8 × 4	312
2 × 1.25	4,000	5 × 3.0	666	8 × 5	250
2 × 1.50	3,333	5 × 3.5	571	8 × 6	208
2 × 1.75	2,857	5 × 4.0	500	8 × 7	178
2 × 2.0	2,500	5 × 4.5	444	8 × 8	156
		5 × 5.0	400		
3 × 0.5	6,666				
3 × 1.0	3,333	6 × 1	1,666	9 × 1	1,111
3 × 1.5	2,222	6 × 2	833	9 × 2	555
3 × 2.0	1,666	6 × 3	555	9 × 3	370
3 × 2.5	1,333	6 × 4	416	9 × 4	277
3 × 3.0	1,111	6 × 5	333	9 × 5	222
		6 × 6	277	9 × 6	185
4 × 0.5	5,000			9 × 7	158
4 × 1.0	2,500	7 × 1	1,428	9 × 8	138
4 × 1.5	1,666	7 × 2	714	9 × 9	123
4 × 2.0	1,250	7 × 3	476	9 × 10	111
4 × 2.5	1,000	7 × 4	357		
4 × 3.0	833	7 × 5	285		
4 × 3.5	744	7 × 6	238	10 × 10	100
4 × 4.0	625	7 × 7	204		

*In order to obtained the number of plants per hectare, divide 10,000 by the product of the spacing in the rows expressed in meters

INDEX

D

E

K

L

M

N

O

P

U

V

W

X

Y